Sustainability of the Food System

Sustainability of the Food System

Sovereignty, Waste, and Nutrients Bioavailability

Edited by

Noelia Betoret, PhD

Universitat Politècnica de València
València, Spain

Ester Betoret, PhD

Instituto de Agroquímica y Tecnología de Alimentos
Consejo Superior de Investigaciones Científicas
Paterna, Spain

ELSEVIER

ACADEMIC PRESS
An imprint of Elsevier

Academic Press is an imprint of Elsevier
125 London Wall, London EC2Y 5AS, United Kingdom
525 B Street, Suite 1650, San Diego, CA 92101, United States
50 Hampshire Street, 5th Floor, Cambridge, MA 02139, United States
The Boulevard, Langford Lane, Kidlington, Oxford OX5 1GB, United Kingdom

Notices

Knowledge and best practice in this field are constantly changing. As new research and experience broaden our understanding, changes in research methods, professional practices, or medical treatment may become necessary.

Practitioners and researchers must always rely on their own experience and knowledge in evaluating and using any information, methods, compounds, or experiments described herein. In using such information or methods they should be mindful of their own safety and the safety of others, including parties for whom they have a professional responsibility.

To the fullest extent of the law, neither the Publisher nor the authors, contributors, or editors, assume any liability for any injury and/or damage to persons or property as a matter of products liability, negligence or otherwise, or from any use or operation of any methods, products, instructions, or ideas contained in the material herein.

Library of Congress Cataloging-in-Publication Data
A catalog record for this book is available from the Library of Congress

British Library Cataloguing-in-Publication Data
A catalogue record for this book is available from the British Library

ISBN: 978-0-12-818293-2

For information on all Academic Press publications visit our website at https://www.elsevier.com/books-and-journals

Publisher: Charlotte Cockle
Acquisitions Editor: Megan Ball
Editorial Project Manager: Devlin Person
Project Manager: Kiruthika Govindaraju
Cover Designer: Alan Studholme

Typeset by TNQ Technologies

Contents

Contributors

C.D. Anagnostopoulos
School of Agriculture, Laboratory of Ecology and Environmental Protection, Aristotle University of Thessaloniki, Thessaloniki, Greece

Lillian Barros
Centro de Investigação de Montanha, CIMO, Instituto Politécnico de Bragança, Campus de Santa Apolónia, Bragança, Portugal

Ester Betoret, PhD
Instituto de Agroquímica y Tecnología de Alimentos, Consejo Superior de Investigaciones Científicas, Paterna, Spain

Noelia Betoret, PhD
Universitat Politècnica de València, València, Spain

Ludmiła Bogacz-Radomska
Adaptive Food Systems Accelerator, Wrocław University of Economics, Wrocław, Poland; Department of Biotechnology and Food Analysis, Wrocław University of Economics Wrocław, Poland

Alfredo Cassano
Institute on Membrane Technology, ITM-CNR, Rende, Cosenza, Italy

A. Chiralt
Instituto de Ingeniería de Alimentos para el desarrollo, Universitat Politècnica de València, Valencia, Spain

S. Collazo
Instituto de Ingeniería de Alimentos para el desarrollo, Universitat Politècnica de València, Valencia, Spain

Carmela Conidi
Institute on Membrane Technology, ITM-CNR, Rende, Cosenza, Italy

M. D'Alessandro
Department of Agricultural and Food Sciences, Campus of Food Science, University of Bologna, Cesena, Italy

Marco Dalla Rosa, PhD
Department of Agricultural and Food Sciences, Alma Mater Studiorum-Università di Bologna, Cesena, Forli - Cesena, Italy; Professor, Department of Agricultural and Food Sciences, Alma Mater Studiorum-Università di Bologna, Cesena, Forli - Cesena, Italy

Isabel C.F.R. Ferreira
Centro de Investigação de Montanha, CIMO, Instituto Politécnico de Bragança, Campus de Santa Apolónia, Bragança, Portugal

Esperanza M. Garcia-Castello
Institute of Food Engineering for Development, IuIAD, Universitat Politècnica de València, Valencia, Spain

C. Gonzalez-Martinez, PhD
Instituto de Ingeniería de Alimentos para el desarrollo, Universitat Politècnica de València, Valencia, Spain

Joanna Harasym, DSc, PhD, MSc
Adaptive Food Systems Accelerator, Wrocław University of Economics, Wrocław, Poland; Department of Biotechnology and Food Analysis, Wrocław University of Economics Wrocław, Poland

E. Hernandez-García
Instituto de Ingeniería de Alimentos para el desarrollo, Universitat Politècnica de València, Valencia, Spain

Urszula Kaim
Adaptive Food Systems Accelerator, Wrocław University of Economics, Wrocław, Poland; Department of Biotechnology and Food Analysis, Wrocław University of Economics Wrocław, Poland

K.L. Kalburtji
School of Agriculture, Laboratory of Agronomy, Aristotle University of Thessaloniki, Thessaloniki, Greece

Andriana Kechagia, MSc
Terrapi World S.L., Ibiza, Spain

R. Lanciotti
Department of Agricultural and Food Sciences, Campus of Food Science, University of Bologna, Cesena, Italy; CIRI - Agrifood (Interdepartmental Centre of Industrial Agrifood Research), University of Bologna, Cesena, Italy

V.D. Litskas
School of Agriculture, Laboratory of Ecology and Environmental Protection, Aristotle University of Thessaloniki, Thessaloniki, Greece; Department of Agricultural Sciences, Biotechnology and Food Science, Cyprus University of Technology, Limassol, Cyprus

A.P. Mamolos
School of Agriculture, Laboratory of Ecology and Environmental Protection, Aristotle University of Thessaloniki, Thessaloniki, Greece

Bruno Melgar
Institute of Food Engineering for Development, IuIAD, Universitat Politècnica de València, Valencia, Spain

G.C. Menexes
School of Agriculture, Laboratory of Agronomy, Aristotle University of Thessaloniki, Thessaloniki, Greece

C. Menzel
Division of Glycoscience, Department of Chemistry, School of Engineering Sciences in Chemistry, Biotechnology and Health, KTH Royal Institute of Technology, AlbaNova University Centre, Stockholm, Sweden

Remigiusz Oledzki
Adaptive Food Systems Accelerator, Wrocław University of Economics, Wrocław, Poland; Department of Biotechnology and Food Analysis, Wrocław University of Economics Wrocław, Poland

F. Patrignani, PhD
Department of Agricultural and Food Sciences, Campus of Food Science, University of Bologna, Cesena, Italy; CIRI - Agrifood (Interdepartmental Centre of Industrial Agrifood Research), University of Bologna, Cesena, Italy

D.P. Platis
School of Agriculture, Laboratory of Ecology and Environmental Protection, Aristotle University of Thessaloniki, Thessaloniki, Greece

Carmen Carla Quiroga Ledezma, PhD
Director, Centro de Investigaciones Agrícolas y Agroindustriales, Universidad Privada Boliviana, Cochabamba, Bolivia

Antonio D. Rodriguez-Lopez
Institute for Industrial, Radiophysical and Environmental Safety, ISIRYM, Universitat Politècnica de València, Valencia, Spain

D. Rodríguez Alberto, B.S., M.S.
PhD Candidate, Golisano Institute for Sustainability, Rochester Institute of Technology, Rochester, NY, United States

Salvador Roig-Coll, MSc
Terrapi World S.L., Ibiza, Spain

M.C. Stavrinides
Department of Agricultural Sciences, Biotechnology and Food Science, Cyprus University of Technology, Limassol, Cyprus

Silvia Tappi
Department of Agricultural and Food Sciences, Alma Mater Studiorum-Università di Bologna, Cesena, Forli - Cesena, Italy

T.A. Trabold, B.S., PhD
Department Head, Golisano Institute for Sustainability, Rochester Institute of Technology, Rochester, NY, United States

A.C. Tsaboula
School of Agriculture, Laboratory of Ecology and Environmental Protection, Aristotle University of Thessaloniki, Thessaloniki, Greece

Urszula Tylewicz
Department of Agricultural and Food Sciences, Alma Mater Studiorum-Università di Bologna, Cesena, Forli - Cesena, Italy

L. Vannini
Department of Agricultural and Food Sciences, Campus of Food Science, University of Bologna, Cesena, Italy; CIRI - Agrifood (Interdepartmental Centre of Industrial Agrifood Research), University of Bologna, Cesena, Italy

PART

I

Food sovereignty

CHAPTER 1

Native food crops for present and future generations: their role in nutrition and health

Carmen Carla Quiroga Ledezma, PhD
Director, Centro de Investigaciones Agrícolas y Agroindustriales, Universidad Privada Boliviana, Cochabamba, Bolivia

1. Introduction

The second Sustainable Development Goal (SDG2) aims at ending hunger and ensuring access to food by all, ending malnutrition in all its forms, increasing agricultural productivity and incomes of small-scale food producers, ensuring sustainable food production systems, and implementing resilient agricultural practices, as well as maintaining genetic diversity (UN, 2019). Even though in the past decades, agricultural productivity has increased, leading to an increase in food, the way in which it was achieved needs to be adapted to reach SDG2, because it has put pressure on natural resources and often resulted in negative outcomes such as land degradation, deforestation, salinization due to irrigation, overexploitation of groundwater, pollution of water bodies and soil caused by agrochemicals, emission of greenhouse gases, built-up resistance to pests, and erosion of biodiversity (FAO, 2017a).

2. Nutrition and health relationship

Lifestyle, dietary patterns, and consumer behavior have been changing in recent years, driven by globalization, urbanization, income growth, and modern food systems (GPAFSN, 2016), which clearly impact malnutrition in all its forms: wasting, stunting, underweight, micronutrient deficiencies, overweight, and obesity, as well as impacting diet-related noncommunicable diseases such as cardiovascular diseases, cancer, chronic respiratory diseases, and diabetes (FAO et al., 2018). In addition, climate variability and extreme environments are negatively affecting food security, as well as reinforcing other underlying causes of malnutrition (GPAFSN, 2016). Pregnant and lactating women, adolescent girls, young children, old people, people who are ill or immune compromised, indigenous people, poor people from rural areas, and rural migrants are more vulnerable to malnutrition (HLPE, 2017).

Sustainability of the Food System. https://doi.org/10.1016/B978-0-12-818293-2.00001-X

Statistics are overwhelming: 821 million people are undernourished; nearly 151 million children under five are stunted, over 50 million are wasted at increased risk of morbidity and mortality, whereas 41 million are overweight or obese; more than 1.9 billion adults are overweight or obese, whereas 462 million are underweight (WHO, 2018a). More than 41 million people die every year due to noncommunicable diseases: over 17 million from cardiovascular diseases, 9 million from cancer, 3.9 million from respiratory diseases, and 1.6 million from diabetes (WHO, 2018b).

To decrease these trends, enough sustainable food should be supplied, and healthy and safe diets, as well as a variety of foods such as vegetables, fruits, whole grains and cereals, dairy foods, and animal- and plant-based proteins, which have been processed to remove antinutrients or components within foods that interfere with nutrient absorption, should be eaten to provide energy requirements and sufficient macro- and micronutrients and bioactive compounds. The consumption of nutrient-poor foods high in energy, saturated and transfats, added sugars, and salt should be limited (HLPE, 2017). Nowadays, food is intended not only to nourish and prevent nutrition-related diseases but also to improve physical and mental well-being.

Food contains major (proteins, lipids, and carbohydrates) and minor (vitamins and minerals) substances, which are necessary to maintain life and allow growth and reproduction. These substances provide energy, contribute to structure, and regulate biological processes; about 45 of these substances are considered essential to human life and must be supplied in the diet. Food also contains bioactive compounds that are phytochemicals, i.e., nonessential, which may have health-promoting properties on the whole body or specific tissues or cells due to their capability to modulate metabolic processes.

3. Native food crops

Twelve crops provide about 75% of the world's food, i.e., 6.7 billion tons per year (FAO, 2017b). Contribution of these crops to agriculture and food production is important; even if they may provide enough calories to prevent hunger, they do not provide all the nutrients necessary to prevent malnutrition (Li et al., 2018). Four of these crops are native to Meso and South America (corn, potato, cassava, and tomato) (FAO, 2017a). These regions are amazingly rich in biodiversity, especially the Andes and the Amazonia (FAO and PAR, 2010), and offer a great potential to rediscover nutritious and healthy local ecotypes and varieties, which were wisely used by ancient cultures and developed for a wide range of temperature, precipitation, ultraviolet irradiation, soil types, and pest and disease resistance, e.g., the Incas cultivated almost as many species of plants as the farmers of all Asia and Europe and grew a wealth of grains, legumes, roots and tubers, vegetables, fruits, and nuts (NRC, 1989).

The neglected and underutilized food crops of the Andean and Amazonian regions, as well as other species that have not been yet domesticated, offer enormous opportunities for fighting poverty, hunger, and malnutrition, locally and worldwide (Li et al., 2018). Many of the food crops continue to be grown by indigenous farmers, whereas others crops grow wild, but they represent unique sources of genetic material to improve and develop novel crops and foods that will meet the actual needs that the human race is facing (FAO, 2019). These species are less demanding of their environment and so offer stronger resilience to climate change and food insecurity. In addition, since they are more resistant to stressful events, their levels of essential minerals, vitamins, and bioactive compounds are higher and their macronutrient profiles of fat, protein, and carbohydrate are also better than those of commercial crops (Tapia and Fries, 2007). Ancient farmers took advantage of natural plant adaptations to extreme environments to domesticate a unique group of crops and combined their cultivation with the use of complex irrigation canals, terraces, crop rotations, intercropping, livestock, cropping techniques, and soil preservation practices (Flores et al., 2003).

Some domesticated, semidomesticated, and wild species from the Andes and the Amazonia are presented, which were selected due to their nutritional and functional properties, as well as their capacity for adaptability to harsh environments. Some of them played an important role in the diet and traditional medicine of ancient cultures and have caught attention worldwide lately. Therefore, some research has been carried out to identify their chemical composition and the effects of those compounds on nutrition and health through in vitro and in vivo trials to verify the attributes traditionally granted to those species; however, there is still much to be done to confirm or refute them. It is worthwhile to bear in mind that sometimes reported data may vary among authors due to several reasons, such as the variety or ecotype, growing environment, harvesting and processing conditions, sample preparation, analytical methods, and data processing and presentation.

4. Andean grains

Andean grains, together with potato, were staple foods of pre-Hispanic cultures. Pollen, seeds and inflorescences samples, and pictorial figures were found in archeological sites dated thousands of years ago.

4.1 *Amaranthus caudatus* Linnaeus

Amaranth's most common local names are kiwicha, millmi, coimi, and qamasa. It grows best at 2800–3300 m a.s.l., adapts to varying environments, and tolerates high temperatures, bright sunlight, and dry conditions because it follows the C4 carbon-fixation pathway. Amaranth is an annual, herbaceous, and slightly bush-like plant, with colorful panicles like quinoa and canihua, and can reach a height of 0.6–3 m at maturity. Seeds (grains) are coated by a shiny layer and surrounded by an embryo and are about 1 mm in diameter, and their color ranges from black through red to the more common ivory or white (Fig. 1.1) (NRC, 1989). Traditionally, the leaves are used as vegetables, like spinach, the red flowers as colorants in beverages, and the seeds as whole grain or flour.

FIGURE 1.1

Andean grain: *Amaranthus caudatus*, *Chenopodium pallidicaule*, *Chenopodium quinoa*, *Lupinus mutabilis*, and *Zea mays*.

Amaranth grain is a good source of protein (12.80%−15.88% fresh weight, $N \times 6.25$), as compared with cereals, with high levels of lysine (5.22−6.04 g/16 g N) and sulfur amino acids but low levels of leucine, which is the limiting amino acid (Gamel et al., 2004). It is also a good source of lipid (6.31% −7.56% fw) as it is rich in β-tocopherol, α-tocopherol, linoleic, oleic, and palmitic acids, and especially in squalene (Bruni et al., 2001). Carbohydrate is the major fraction in the grain (55.5%−63.7% fw), although fiber (2.68%−7.49% fw) seems to be low; insoluble fiber is composed of homogalacturonans and rhamnogalacturonan-I with arabinan side chains, as well as highly branched xyloglucans and cellulose, and soluble fiber of xyloglucans and arabinose-rich pectic polysaccharides (Lamothe et al., 2015). The grain contains phenolic acids (16.8−32.9 mg/100 g) such as caffeic, ferulic, *p*-coumaric, *p*-OH-benzoic, vanillic, sinapic, and protocatechuic acids; the last two acids are not present in canihua or quinoa. Flavonoids are not detectable, only traces of quercetin. β-cyanins are present in amaranth's seedlings, leaves, and inflorescences (Repo-Carrasco-Valencia et al., 2010). It is also a rich source of minerals and trace elements (Nascimento et al., 2014). Several saponins have been identified in amaranth grain (Rastrelli et al., 1995), as well as lectin (amaranthin) (Rinderle et al., 1989).

4.2 *Chenopodium pallidicaule* Aellen

Canihua's most common native names are cañahua, kañawa, cañihua, and kañiwa. It grows best at 3800−4200 m a.s.l., flourishes in the extreme highland environment where other grain crops grow unreliably or not at all, and is resistant to a combination of frost, drought, salt, and pests. It loses most of its grains by dehiscence, which indicates that this spiecies is not yet completely domesticated. Canihua is an annual, erect, or semiprostrate weed, which reaches a height of 20–70 cm at maturity. The plant's architecture is designed to support the Altiplano's harsh cold (NRC, 1989). Seeds are coated by the perigonium, which is usually gray in color, and are surrounded by the embryos, which are about 1–1.2 mm in diameter, and their color ranges from black to dark or light brown (Fig. 1.1); they contain little or no saponins; however, harvesting and decoating them are laborious (Quiroga Ledezma et al., 2018). Edible parts of canihua, leaves, flowers, and seeds are used in the same manner as the amaranth's.

Canihua grain is a good source of protein (13.29%–18.28% fw, $N \times 6.25$); its content is higher than that of amaranth and quinoa and is primarily composed of albumins and globulins. Its quality is comparable with that of whole dried milk, although threonine is the limiting amino acid (White et al., 1955). It is also a good source of lipid (4.46%–7.92% fw) as it is rich in unsaturated fatty acids (~72%); the sterol fraction is abundant mainly in δ7-stigmasterol and δ7,22-stigmastedienol (Gallego Villa et al., 2014) and the lipid fraction in γ-tocopherol and α-tocopherol. Carbohydrate is the major fraction in the grain (52.4%–61.5% fw), whereas fiber (5.33%–7.52% fw) content is not (Repo-Carrasco-Valencia et al., 2010). The grain contains phenolic acids (18.3–40.1 mg/100 g) such as caffeic, ferulic, *p*-coumaric, *p*-OH-benzoic, and vanillic acids. It also contains flavonoids (46.1–144.3 mg/100 g) such as, to a larger extent, quercetin and isorhamnetin, whereas myricetin, kaempferol, and rhamnetin are found in some varieties (Repo-Carrasco-Valencia et al., 2010), as well as other phenolic compounds such as catechin, catechin gallate, resorcinol, and *p*-methylresorcinol (Peñarrieta et al., 2008). It is also a source of vitamins, minerals, and trace elements (De Bruni, 1964). Several saponins have been identified in canihua grain (Rastrelli et al., 1996).

4.3 *Chenopodium quinoa* Willdenow

Quinoa's most common native names are kiuna, quinua, jupha, and suba, and the Incas considered it sacred. It grows at 0–4000 m a.s.l. and adapts to different environmental conditions, e.g., quinoa races cultivated in the surroundings of salt flats are resistant to extreme xerophyte conditions (Quiroga, 2018). Quinoa is an annual, broad-leaved herb that reaches a height of about 1–2 m at maturity (Fig. 1.1). Seeds of most varieties and ecotypes contain saponins in the outer layers of the episperm, which are bitter and must be removed before human consumption; seeds are also surrounded by the embryo, are about 2 mm in diameter, and their color ranges from almost white through pink, orange or red to brown or black (Quiroga et al., 2015). Parts of quinoa that are consumed are like those of the amaranth and canihua.

Quinoa grain is one of the best sources of protein (11.32%–14.72% fw, $N \times 6.25$) in the vegetable kingdom of exceptional high quality. It is particularly rich in essential amino acids, such as sulfur amino acids, lysine, and aromatic amino acids; therefore, its PER values are comparable with casein. Albumins and globulins are the major protein fraction, and prolamins are the minor protein fraction (Koziol, 1992); it is also a good source of lipid (3.95%–6.85% fw) with a high content of linoleic and γ-linolenic acids, γ-tocopherol and α-tocopherol (Tang et al., 2015). Carbohydrate is the major fraction in the grain (59.4%–70.8% fw), and its fiber (8.8% –14.1% fw) resembles that of fruits, vegetables, and legumes. The insoluble fraction (78%) is composed mainly of galacturonic acid, arabinose, galactose, xylose, and glucose subunits, whereas the soluble fraction (22%) is composed mostly of glucose, galacturonic acid, and arabinose units (Graf et al., 2015). The grain contains high amounts of phenolic acids (30.3–59.7 mg/100 g), which exist in free, but also in conjugated forms; among them, vanillic and ferulic acids and their derivatives are the highest phenolic acids, and quercetin and kaempferol and their glycosides are the main flavonoids. It also comprises carotenoids, mostly translutein and zeaxanthin (Tang and Tsao, 2017); betacyanins (betanin, isobetanin, and amaranthin) in red-violet varieties; and betaxanthins in yellow-orange varieties (Escribano et al., 2017). It is rich in micronutrients, such as vitamins and minerals, as well as saponins, phytosterols, and phytoecdysteroids (Navruz-Varli et al., 2016).

4.4 *Lupinus mutabilis* Sweet

Tarwi's most common local names are tarwi, tauri, and chocho. It grows best at 2700–3800 m a.s.l. and tolerates frost, drought, a wide range of soils, and many pests. Tarwi is an erect annual plant with multicolored purple to blue flowers (Fig. 1.1) and reaches a height of 0.5–2.5 m at maturity. Its root presents symbiotic nodules with bacteria of the genus *Rhizobium lupini* capable of fixing nitrogen gas from the atmosphere and benefitting the soil in which it grows (Tapia and Fries, 2007), and its seeds are edible. Pods contain 4–12 ovoid seeds of 0.5–1.5 cm across, which are covered by a hardened tegument that represents 10% of its weight, and they are bitter due to the presence of alkaloids, which must be removed prior to human consumption (NRC, 1989).

Tarwi grain is as rich, or richer, in protein (~38.9% fw, $N \times 6.25$) than other legumes and other lupin species, and its protein is characterized by a low level of sulfur amino acids, a poor PER, and a high serum urea concentration; however, the apparent digestibility (80.4%) is superior to that of most other plant proteins. The essential amino acid profile among lupine species varies little, but tarwi has the highest lysine content (Carvajal-Larenas et al., 2016). One of the tarwi proteins, γ-conglutin, has attracted attention due to its binding capacity to some small regulatory proteins, including insulin (Muñoz et al., 2018). The grain is also a good source of lipid (~18.9% fw) and, together with protein, represents more than 60% of the dry seed; it is rich in unsaturated (79.38%) fatty acids and has a high content of essential fatty acids, such as linoleic and linolenic acids, as well as γ-tocopherol and δ-tocopherol (Villacrés et al., 2010). Carbohydrate (~23.81% fw) content is low as compared with the grains already presented, although its fiber (~6.13% fw) content is comparable; it has a high content of the indigestible α-galactoside (~10%) and very low content of sucrose (~1.5%) as the only detectable digestible sugar (Gross et al., 1989). The grain is also a source of minerals, mainly calcium, and, to a lesser degree, vitamins. In this particular species, the alkaloid content is high, although there are already some varieties with a lower content of alkaloids.

4.5 *Zea mays* Linnaeus

Purple corn's usual local name is kulli sara. It grows best at 1800–2800 m a.s.l. and adapts to different environments, and although it is sensitive to frost, it is tolerant to pests. Purple corn is an annual plant and reaches a height of 1.8–2.5 m at maturity (INIA, 2019), with leaves originate in the knots and from some of whose axils the spike is born, known as the ear, which includes the central axis where the flowers that will give origin to the seeds (kernels) are inserted, which weigh about 2.93–4.25 g (Fig. 1.1) (Salinas Moreno et al., 2013).

In general, corn kernel is composed of starch (70%–75% dry weight), which comprises most of the total kernel (~80% dw), with high fiber content, mainly consisting of hemicellulose, cellulose, and lignin, but there is also β-glucan and arabinoxylan present (Singh et al., 2019). Protein (8%–10% dw, $N \times 6.25$) is composed mainly of prolamins, which are collectively referred to as zeins, with relatively poor nutritional quality due to the low content of lysine and tryptophan. Lipid (4%–5% dw) is composed mostly of triacylglycerides, which contain a mixture of saturated and unsaturated fatty acids such as linolenic, oleic, palmitic, and stearic acids (Boyer and Hannah, 2001). Immature kernel contains relatively high levels of sugars and lower amounts of starch, protein, and lipid. It is also a source of minerals, although the content is not as high as that of grains already presented, and it is a rich source of natural antioxidants (Nascimento et al., 2014). Phenolic compounds are present in the kernel, including mainly phenolic acids such as *p*-coumaric, vanillic, protocatechuic, chlorogenic, caffeic, rutin, and ferulic acids, and flavonoids such as quercetin, morin, naringenin, kaempferol, and hesperitin derivatives (Pedreschi and Cisneros-Zevallos, 2007) (Ramos-Escudero et al., 2012). Anthocyanins, which are also found in the kernel, comprise cyanidin, pelargonidin, and peonidin. Additional variation of the anthocyanins is derived from different types, combinations, and degree of hydroxylation, glucosylation, methylation, and acylation (Cuevas Montilla et al., 2011) (Lao et al., 2017). Purple corn produces anthocyanins throughout the plant, with an especially high content in the husk and cob, and lower in the kernel (Li et al., 2008).

5. Andean roots and tubers

Pre-Hispanic cultures domesticated about 17 species of roots and tubers in the Andes, making this the largest known geographical concentration of underground food crops worldwide; the genetic diversity and potential of these species is extraordinary.

5.1 *Lepidium peruvianum* Chacón

This scientific name replaces that of *Lepidium meyenii*, described by Walpers (1843), which corresponds to a very prostrate wild species (Meissner et al., 2015). Although it was used widely in the past, it was declared, around the 1980s, to be in danger of extinction (NRC, 1989). It grows best at 3900–4200 m a.s.l. and is probably the only food crop worldwide capable of flourishing at an intense ultraviolet radiation, strong winds, very cold climates, and barren and rocky soils (Flores et al., 2003). Maca, the native name of this species, is a herbaceous perennial plant, grown annually, with branched, mostly decumbent stems, and fleshy, starchy tuberous roots (hypocotyl roots), which are edible, as are its leaves. The hypocotyl roots are yellow, yellow with purple bands, purple, and black; attain a size of up to 8 cm in diameter; and have a radish-like shape, and a sweet, tangy flavor (Fig. 1.2). Traditionally, fresh hypocotyl roots are baked, roasted in ashes, or sun-dried, but before eaten, they need to be boiled in water to obtain a soft texture (NRC, 1989).

Dry hypocotyl root proximal composition is similar to some cereals and better than potato, which is composed of protein ($\sim$10.2% dw, $N \times 6.25$), lipid ($\sim$2.2% dw), carbohydrate ($\sim$59.0% dw), fiber ($\sim$8.5% dw), and ash ($\sim$4.9% dw). The fatty acid fraction is rich in linoleic, palmitic, and oleic acids, and other polyunsaturated fatty acids are known by the name of macaenes and macamides (benzylalkamides) (Gunstone et al., 1977). The sterol fraction is rich in sitosterol, campesterol, ergosterol, brassicasterol, and δ-7,22-ergostadienol (Dini et al., 1994). Macamides are a result of traditional Andean postharvest drying practices or industrial flaking and oven drying (Esparza et al., 2015). Glucosinolates exist in the fresh or dry hypocotyl root, and the main compounds are the aromatic type such as benzyl glucosinolate and *p*-methoxybenzyl glucosinolate (Li et al., 2001). As a result of the reaction between glucosinolates and the endogenous enzyme myrosinase, isothiocyanates are formed, of which the most important compound is benzyl isothiocyanate (Li et al., 2001). Other compounds are also present in maca hypocotyl roots, such as uridine, malic acid, prostaglandins, flavonoids, anthocyanins, alkaloids, and saponins (Valerio Jr. and Gonzales, 2005).

FIGURE 1.2

Andean roots and tubers: *Lepidum peruvianum*, *Smallanthus sonchifolius*, and *Tropaeolum tuberosum*.

5.2 *Smallanthus sonchifolius* (Poepping and Endlicher) H. Robinson

Polymnia sonchifolia and *Polymnia edulis* are synonyms of this species. Yacon's most common local names are yacón, llakuma, and aricoma. It grows best at 500–3200 m a.s.l. and adapts to seasonal cycles of drought or cold and to poor soil. Yacon is a compact, herbaceous plant with dark-green celery-like leaves and reaches a height of 2 m at maturity; its leaves and tubers are edible. The tubers are irregularly spindle-shaped to round and can vary considerably in shape, size, and sweetness; on the outside, they are tan to purplish brown, but inside they are white, yellow, purple, orange, or yellow, sometimes with magenta dots (Fig. 1.2). A tuber usually weighs 200–500 g but can reach 2 kg (NRC, 1989).

Yacon is considered a highly pest- and pathogen-resistant plant due to the presence of antimicrobial and insecticidal compounds in the leaves, such as kaurenoids and melampolide-type sesquiterpene lactones, being 8β-methacryloyloxymelampolid-14-oic acid methyl ester and fluctuanin with the highest antibacterial activity, although there are also other melampolides such as sonchifolin, uvedalin, and enhydrin (Lin et al., 2003). Its leaves are also rich in phenolic acids such as protocatechuic, chlorogenic, caffeic, and ferulic (Valentova et al., 2003) as well as dicaffeoylquinic acids and their derivatives (Genta et al., 2010).

The fresh tuber is composed mainly of water (69%–83%) and sugar (~20%); dry tuber of protein (6%–7% dw, $N \times 6.25$), lipid (0.4%–1.3% dw), and ash (4%–7% dw) with high levels of potassium and sodium, fiber (4%–6% dw) and simple sugars (~65% dw) (NRC, 1989) such as glucose, fructose, and saccharose as well as β-(2-1)-fructooligosaccharides (inulin-type oligofructans) (Ohyama et al., 1990). The sugar content fluctuates during the growth cycle and following harvest, and among genotypes (Lachman et al., 2004). Resistant starch (19.7%–75.9% dw) content is inversely correlated to fructooligosaccharides (FOS) (6.4%–65% dw) content (Campos et al., 2012). The FOS ($DP < 9$, average $DP = 4.8$) are neither hydrolyzed nor absorbed in the upper part of the gastrointestinal tract but are fermented by *Bifidobacteri* and *Lactobacilli* in the colon, which results in the production of short-chain fatty acids such as propionic, butyric, and acetic acids (Pedreschi et al., 2003). Chlorogenic acid and tryptophan are the major antioxidants (Yan et al., 1999), but they also contain caffeic acid and its derivatives such as esters with the hydroxyl groups of aldaric and altraric acids (Takenaka et al., 2003) and flavonoids such as quercetin (Simonovska et al., 2003).

5.3 *Tropaeolum tuberosum* Ruíz and Pavón

The most common native names are mashua, añu, isaño, and issanu, and it grows best at 2000–4000 m a.s.l. Among Andean tubers, it is one of the highest yielding, easiest to grow, and most resistant to cold and also repels insects, nematodes, and pathogens (Lim, 2016). Mashua habit is prostrate or climbing and reaches about 1 m in diameter and 0.5 m in height with flowers in orange and scarlet color; its tender young leaves, flowers, and tubers are edible, but they are usually boiled before human consumption due to their sharp flavor. The tubers have elongated and deep buds, are conical or ellipsoidal in shape, and have different colors such as white, yellow, orange, red, or purple (Fig. 1.2) (Hind, 2010).

Mashua is quite nutritious for a root crop because it contains high content of protein (7.22%–13.99% dw, $N \times 6.25$), which is biologically incomplete, and lipid (~4.61% dw). Carbohydrate (~75.40% dw) that comprises equal amounts of starch and sugars, fiber (~5.86% dw), and ash (~4.81% dw) contents are similar to that of other Andean roots, although vitamin C and β-carotene contents are higher (Espín et al., 2004). The main glucosinolates in this species are aromatic type; *p*-methoxybenzyl glucosinolate is the most important, but there are also others present, such as *p*-hydroxybenzyl, benzyl, and *m*-methoxybenzyl glucosinales. The products of an enzymatic hydrolysis of glucosinolates are isothiocyanates such as *p*-methoxybenzyl and benzyl isothiocyanates as the major constituents (Ortega et al., 2006). The major anthocyanins are delphinidin di- and triglucosides acylated with acetic acid, and the minor compounds are cyanidins and pelargonidins. Other phenolic compounds are also present in mashua tuber such as gallocatechin, epicatechin, epigallocatechin, procyanidin B_2, proanthocyanidin, quercetin; the derivatives of rutin, hydroxycinnamic, and hydroxybenzoic; and the acids of gallic, *p*-coumaric, *o*-coumaric, cinnamic, and protocatechuic, and rutin quercetin (Chirinos et al., 2008). The content of phenolic compounds depends on the genotype, maturity stage, and postharvest storage, e.g., purple-colored tubers present several times higher antioxidant activity than the yellow-colored tubers (Campos et al., 2019).

6. Andean and Amazonian fruits

Fruits were also important for pre-Hispanic cultures, despite not having such a wide variety of species as Andean grains, roots, and tubers, at least not like the Amazonia, which shelters an amazingly huge diversity of species.

6.1 *Annona muricata* Linnaeus

Soursop's most common local names are sinini, graviola, and guanábana. It grows in tropical and subtropical areas and is a small, slender, evergreen tree that reaches a height of 4–8 m at maturity. It produces an oval, conical, or heart-shaped fruit that is dark green when unripe and a slightly lighter green when ripe, with a weight of 0.9–10 kg. The fruit is a white, cottony-fibrous, juicy flesh with a flavor similar to that of a mixture of pineapple and mango, and it has black seeds scattered throughout the pulp, which become dark brown later (Fig. 1.3) (Pinto et al., 2005).

The white edible pulp corresponds to about 67.5% of the whole soursop fruit, which is composed of water (~82.8% fw), protein (~1.0% fw, $N \times 6.25$) that is biologically incomplete, lipid (~1.0% fw), carbohydrate (~14.6% fw) that has a high content of sugars (12%) such as sucrose, glucose, and fructose, as well as

FIGURE 1.3

Andean and amazonian fruits: *Annona muricata*, *Euterpe oleracea*, *Myrciaria dubia*, *Physalis peruviana*, *Plukenetia volubilis*, and *Solanum betaceum*.

pectin, fiber (~0.8% fw), ash (~0.9% fw) rich in potassium and sodium, and vitamins such as A, B-complex, and C (Badrie and Schauss, 2010). The fruit is a source of cinnamic acid derivatives and *p*-coumaric acid together with several several other phenolic compounds (Jimenez et al., 2014). It contains annonaceous acetogenin compounds such as epomusenin-A and epomusenin-B, epomurinin-A and epomurinin-B, *cis*-annoreticuin, muricins J, K, and L, as well as alkaloids such as annonaine, nornuciferine, and asimilobine (Moghadamtousi et al., 2015). The characteristic flavor of this fruit has been related to many volatile compounds such as esters, which are also present (MacLeod and Pieris, 1981).

The seeds are rich in oil, protein, and minerals, but they also have toxic compounds; together with the leaves and roots, they are also rich in phytonutrients, which is perhaps why they are used in traditional medicine (Vijayalakshmi & Rajeswari V, 2012).

6.2 *Euterpe oleracea* Martius

Asaí or açaí palm tree grows up to 25 m in height, mainly in floodplains, swamps, and upland regions of the Amazonian region. Although the soft interior stem may be used as a source for heart of palm, it is better known for its fruit (Ulbricht et al., 2012). The fruit is round-shaped drupe with a diameter of about 12 mm clustered into bunches; it has one large seed of 7−10 mm that represents around 80%−90% of its volume (Heinrich et al., 2011); the hard, purple epicarp is a very thin layer, and the mesocarp is 1−2 mm in thickness; in addition, ripe fruit is covered with a wax cuticle. The fruit color goes from green to dark purple during the ripening process (Fig. 1.3) (Dias et al., 2012).

Asaí pulp contains biologically active protein (~8.0% dw, $N \times 6.25$) and lipid (~32.5% dw) with oleic acid as the predominant fatty acid, followed by palmitic and linoleic acids, carbohydrate (~52.2% dw), fiber (~44.2% dw), and ash (~3.8% dw) (Schauss et al., 2006) rich in essential minerals such as calcium, iron, magnesium, and zinc; however, the levels of copper and, in particular, manganese are much higher, which leads to a special concern, since iron absorption is impaired by manganese (da Silva et al., 2014). It is a source of carotenoids, lutein, α-carotene, and β-carotene as the major components, but also of vitamins such as vitamin E with higher content of α-tocopherol (Darnet et al., 2011). Asaí fruit has a very high content of phenolic compounds such as gallic and benzoic acids, flavonoids such as taxifolin 3-rhamnoside, luteolin 8-glucoside, apigenin 6-glucoside and catechin, and stilbenes such as resveratrol (Pacheco-Palencia et al., 2009). It also has a remarkably high content of anthocyanins such as cyanidin 3-glucoside and cyanidin 3-rutinoside as the major constituents (Del Pozo-Insfran et al., 2004). It is also a source of sterols such as β-sitosterol (Heinrich et al., 2011) and lignans (Chin et al., 2008).

6.3 *Myrciaria dubia* (H.B.K.) McVaugh

Camu-camu is a deeply rooted shrub that grows in swampy or flooded areas in the Amazon region. It can adapt to different edaphoclimatic conditions and reaches a height of about 4–8 m. The fruit is round with a diameter of about 1.0–3.2 cm, containing three to four seeds. Its pulp is lightly yellowish, but the skin has an intense color ranging from dark red to purple-black at maturity. It has a mild characteristic flavor, but it is extremely acidic, so it is mainly consumed as an ingredient in prepared foods (Fig. 1.3) (Rodrigues et al., 2001).

Camu-camu pulp contains water (94.1%–94.4% fw), protein (0.4%–0.5% fw, $N \times 6.25$) with a high content of essential amino acids such as valine and leucine as compared with other fruits (Zapata and Dufour, 1993), lipid (0.2%–0.3% fw), carbohydrate (3.5%–4.7% fw) with fructose and glucose as the main sugars, fiber (0.1%–0.6% fw), and ash (0.2%–0.3% fw) with high content of minerals such as potassium, calcium, magnesium, and sodium (Castro et al., 2018). It has the highest known content of natural vitamin C (2004.7–2605.8 mg/100 g fruit) (Camargo Neves et al., 2015). Camu-camu is a good source of carotenoids such as lutein, β-carotene, violaxanthin, and luteoxanthin as the major compounds (Zanatta and Mercadante, 2007). It is also a good source of anthocyanin, of which cyanidin 3-glucoside and delphinidin 3-glucoside are its major constituents (Akter et al., 2011), as well as other phenolic compounds such as catechin, quercetin, rutin, myricetin, kaempferol, naringenin, eriodictyol, gallic acid and its derivatives, ellagic acid and its derivatives, and ellagitannins (Fracassetti et al., 2013) (Chirinos et al., 2010). Terpenic compounds such as α-pinene and *d*-limonene are predominant among camu-camu volatile compounds (Franco and Shibamoto, 2000). The seeds are also a source of betulinic acid and other active compounds (Yazawa et al., 2011).

6.4 *Physalis peruviana* Linnaeus

Its most common local names are uchuva, uvilla, and aguaymanto. Goldenberry (cape gooseberry) grows best at 1500–3000 m a.s.l. and is an herbaceous, erect shrub of 45–90 cm height at maturity, perennial in the tropics, and annual in the temperate zones (Fischer and Herrera, 2011). The fruit is encased in a husk that starts out soft and green when young but becomes tough, brown, and paper-like when it reaches maturity and is spherical with a diameter of 1.25–2.5 cm and a weight of 4–10 g, with smooth, glossy orange skin and a juicy pulp containing many very small edible seeds of about 2 mm of diameter, with a pleasant tropical flavor (Fig. 1.3). Each plant produces about 150–300 fruits that are eaten fresh or cooked (Fischer, 2000).

Goldenberry pulp comprises water (~82.8% fw), protein (~0.05–0.3% fw, $N \times 6.25$), lipid (~0.15–0.2% fw) with high proportion of polyunsaturated fatty acids such as γ-linolenic, carbohydrate (~19.6% fw) with sucrose and fructose as the main sugars, fiber (~4.9% fw), and ash (~1.0% fw) rich in phosphorus, iron, zinc, and copper (Ramadan, 2011). It has a high content of vitamin C, but also vitamin A, mostly β-carotene, vitamin B-complex, chiefly thiamine, niacin and riboflavin, vitamin E, mainly γ-tocopherol and α-tocopherol, and vitamin K (Ramadan Hassanien, 2011). It is a rich source of phenolic compounds with quercetin as the main phenolic compound, followed by myricetin and kaempferol; there are also phenolic acids present, such as caffeic, chlorogenic, ferulic, *p*-coumaric, and gallic acids (Puente et al., 2011). It also contains phytosterols such as campesterol, β-sitosterol, and δ5-avenasterol as the most abundant (Ramadan Hassanien, 2011). It even contains other compounds with important biological activity such as physalins A, B, D, and F and glycosides (Wu et al., 2009), and withanolides that are a group of steroidal lactones, withanolide E and 4β-hydroxywithanolide E (Lan et al., 2009).

In traditional medicine, in addition to fruits, leaves and calyxes are also used (Franco et al., 2007).

6.5 *Plukenetia volubilis* Linnaeus

Inca peanut is commonly known as sacha inchi or Inca inchi. It is a perennial, climbing shrub of about 2 m in height, and it grows at 100–2000 m a.s.l. in the rain tropical forests of the Andean region. It produces star-shaped green fruits that turn blackish brown as the fruit matures, yielding edible dark brown seeds, slightly enlarged in the center and squashed toward the edges; the seeds are ovals of 1.5–2 cm in diameter and weigh about 48–100 g (Fig. 1.3) (Krivankova et al., 2012).

Inca peanut seed comprises protein (~24.7% dw, $N \times 6.25$), lipid (~42.0% dw), carbohydrate (~30.9% dw), and ash (~4.0% dw) rich in potassium, magnesium, and calcium (Gutiérrez et al., 2011). Inca peanut protein comprises albumins (~43.7%), globulins (~27.3%), prolamins (~3.0%), and glutelins (~31.9%) (Sathe et al., 2012); it has a good amino acid profile with high contents of cysteine, tyrosine, threonine, and tryptophan, although lysine and leucine are its limiting essential amino acids; its quality is comparable with soy protein and is higher than other oilseed proteins (Hamaker et al., 1992). Its oil has a high degree of unsaturation and contains high levels of linoleic and linolenic acids, representing about 86% of the total fatty acids, which is comparable with linseed oil (Fanali et al., 2011). Phenyl alcohol, flavonoid, secoridoid, and lignan-type phenolics are also present in Inca peanut seed oil (Wang et al., 2018). It is a good source of tocopherols such as γ-tocopherol and δ-tocopherol; phytosterols such as β-sitosterol, stigmasterol, and campesterol; and carotenoids (Chirinos et al., 2013).

The leaf is also a source of biological active compounds and is therefore used in traditional medicine (Lima Nascimento et al., 2013).

6.6 *Solanum betaceum* (Cavanilles) Sendtner

Cyphomandra betacea is a synonym of this species, and the most common native names are tomate de árbol and tomate de monte. Tree tomato, or tamarillo, in its wild form, thrives in climates of humid mountain forests (Buono et al., 2018), grows best at 1500–3200 m a.s.l. in the Andean region, and is a shrub of 3–4 m high, with gray bark and perennial foliage; its fruit are edible raw or cooked. The fruit is an ovoid-apiculate berry that is green when it is immature and yellow, orange, red, or purple when it is ripe, has 4.0–8.0 cm length and 4.0–6.0 cm width, weighs about 30.5–154.2 g, and contains 3.4–282.2 seeds. (Fig. 1.3). The fruit's skin is thin, smooth, and resistant, and its cuticle has a bitter taste; the pulp is juicy, with a flavorful sweet and sour taste (Acosta-Quezada et al., 2011).

The edible part of tree tomato fruit is about 85% and comprises water (~87.8% fw), protein (~1.5% fw, $N \times 6.25$), lipid (~1.3% fw), carbohydrate (~10.3% fw) with sucrose as the main sugar, fiber (~4.2% fw), and ash (~1.0% fw) rich in potassium and iron (Carrillo-Perdomo et al., 2015). It contains interesting types of polysaccharides, a highly methoxylated homogalacturonans mixed with type I arabinogalactans, a linear (1 → 5)-linked α-L-arabinan, and a linear (1 → 4)-β-D-xylan (Erdmann do Nascimento et al., 2016). It is a source of vitamin C and a good source of carotenoids, which are mainly present as esterified forms such as β-carotene, β-cryptoxanthin, zeaxanthin, and lutein (Mertz et al., 2010). The yellow variety lacks anthocyanins, but the purple-red variety has a high content of anthocyanins such as pelargonidin rutinoside, delphinidin rutinoside, cyanidin rutinoside, and delphinidin glucosyl rutinoside. In addition, hydroxycinnamic acids are also present with the major compounds of caffeoylquinic and dicaffeoylquinic acids (Mertz et al., 2009).

7. Conclusions

The species that were presented in this chapter have exceptional nutritional and functional properties. They have a high content of essential amino acids, fatty acids, and minerals, as well as polypeptides, vitamins, carotenoids, and bioactive compounds such as phenolic acids, polyphenols, flavonoids, flavonols, and tannins, which are related to health benefits and risk reduction of certain diseases.

The work done by the natives of the Andes and the Amazonia such as improvement of the species' nutritional and functional properties, yield, capacity to adapt to the environmental conditions of these regions—sometimes even to extreme conditions such as the Altiplano's—was important because they inherited a great germplasm collection, which can be used for further improvement to solve problems related to food safety and sovereignty, and climate change. Many countries worldwide have already bet on these species, and there are more and more research projects to prove or discover their nutritional and health benefits. There is still much to be done, but the future could be promising if they are taken into account.

References

Acosta-Quezada, P.G., Martínez-Laborde, J.B., Prohens, J., 2011. Variation among tree tomato (*Solanum betaceum* Cav.) accessions from different cultivar groups: implications for conservation of genetic resources and breeding. Genetic Resources and Crop Evolution 58 (6), 943–960.

Akter, M.S., Oh, S., Eun, J.-B., Ahmed, M., 2011. Nutritional compositions and health promoting phytochemicals of camu-camu (*Myrciaria dubia*) fruit: a review. Food Research International 44, 1728–1732.

Badrie, N., Schauss, A.G., 2010. Soursop (*Annona muricata* L.): composition, nutritional value, medicinal uses, and toxicology. In: Watson, R.R., Preedy, V.R. (Eds.), Bioactive Foods in Promoting Health: Fruits and Vegetables. Academic Press, pp. 621–643.

Boyer, C.D., Hannah, L.C., 2001. Kernel mutants of corn. In: Hallauer, A.R. (Ed.), Specialty Corns. CRC Press LLC, Boca Raton, pp. 1–31.

Bruni, R., et al., 2001. Wild *Amaranthus caudatus* seed oil, a nutraceutical resource from Ecuadorian Flora. Journal of Agriculture and Food Chemistry 49, 5455–5460.

Buono, S., et al., 2018. Solanum Betaceum (Cav). Sendt. Tomate Árbol, first ed. IICA.

Camargo Neves, L., et al., 2015. Bioactive compounds and antioxidant activity in pre-harvest camu-camu [*Myrciaria dubia* (H.B.K.) McVaugh] fruits. Scientia Horticulturae 186, 223–229.

Campos, D., et al., 2012. Prebiotic effects of yacon (*Smallanthus sonchifolius* Poepp. & Endl), a source of fructooligosaccharides and phenolic compounds with antioxidant activity. Food Chemistry 135, 1592–1599.

Campos, D., et al., 2019. Postharvest storage and cooking techniques affect the stability of glucosinolates and myrosinase activity of Andean mashua tubers (*Tropaeolum tuberosum*). International Journal of Food Science and Technology 1–9.

Carrillo-Perdomo, E., et al., 2015. Andean berries from Ecuador: a review on botany, agronomy, chemistry and health potential. Journal of Berry Research 5, 49–69.

Carvajal-Larenas, F.E., et al., 2016. *Lupinus mutabilis*: composition, uses, toxicology, and debittering. Critical Reviews in Food Science and Nutrition 56, 1454–1487.

Castro, J.C., Maddox, J.D., Iman, S.A., 2018. Camu-camu-*Myrciaria dubia* (Kunth) McVaugh. In: Rodrigues, S., de Oliveira Silva, E., Edy, S.d.B. (Eds.), Exotic Fruits Reference Guide. Academic Press, pp. 97–105.

Chin, Y.-W., Chai, H.-B., Keller, W.J., Kinghorn, A.D., 2008. Lignans and other constituents of the fruits of *Euterpe oleracea* (açai) with antioxidant and cytoprotective activities. Journal of Agricultural and Food Chemistry 56, 7759–7764.
Chirinos, R., et al., 2008. Phenolic profiles of andean mashua (*Tropaeolum tuberosum* Ruiz & Pavon) tubers: identification by HPLC-DAD and evaluation of their antioxidant activity. Analytical, Nutritional and Clinical Methods 106, 1285–1298.
Chirinos, R., et al., 2010. Antioxidant compounds and antioxidant capacity of Peruvian camu camu (*Myrciaria dubia* (H.B.K.) McVaugh) fruit at different maturity stages. Food Chemistry 120, 1019–1024.
Chirinos, R., et al., 2013. Sacha inchi (*Plukenetia volubilis*): a seed source of polyunsaturated fatty acids, tocopherols, phytosterols, phenolic compounds and antioxidant capacity. Food Chemistry 141, 1732–1739.
Cuevas Montilla, E., Hillebrand, S., Antezana, A., Winterhalter, P., 2011. Soluble and bound phenolic compounds in different Bolivian purple corn (*Zea mays* L.) cultivars. Journal of Agriculture and Food Chemistry 7068–7074, 59.
da Silva Santos, V., et al., 2014. Anthocyanin=Rich Açaí (Euterpe oleracea Mart.) Extract attenuates manganese-induced oxidative stress in rat primary astrocyte cultures. Journal of Toxicology and Environmental Health, Part A 77, 390–404.
Darnet, S., Serra, J.I., da Cruz Rodrigues, A.M., Meller da Silva, L.H., 2011. A high-performance liquid chromatography method to measure tocopherols in assai pulp (Euterpe oleracea). Food Research International 44, 2107–2111.
De Bruni, A., 1964. Investigation of the food value of quinua and cañihua seed. Journal of Food Science 29 (6), 872–876.
Del Pozo-Insfran, D., Brenes, C.H., Talcott, S.T., 2004. Phytochemical composition and pigment stability of acai (*Euterpe olerace* Mart.). Journal of Agricultural and Food Chemistry 52, 1539–1545.
Dias, A.L.S., et al., 2012. A rapid validated UHPLC-PDA method for anthocyanins quantification from Euterpe oleracea fruits. Journal of Chromatography B 907, 108–116.
Dini, A., et al., 1994. Chemical composition of *Lepidium meyenii*. Food Chemistry 49, 347–349.
Erdmann do Nascimento, G., Iacomini, M., Cordeiro, L.M.C., 2016. A comparative study of mucilage and pulp polysaccharides from tamarillo fruit (*Solanum betaceum* Cav.). Plant Physiology and Biochemistry 104, 278–283.
Escribano, J., et al., 2017. Characterization of betalains, saponins and antioxidant power in differently colored quinoa (*Chenopodium quinoa*) varieties. Food Chemistry 234, 285–294.
Esparza, E., et al., 2015. Bioactive maca (*Lepidium meyenii*) alkamides are a result of traditional Andean postharvest drying practices. Phytochemistry 116, 138–148.
Espín, S., Villacrés, E., Brito, B., 2004. Caracterizacion fisico-Quimica, nutricional y funcional de raices y tuberculos andinos. In: Barrera, V.H., Tapia, C.G., Monteros, A.R. (Eds.), Raices y Tuberculos Andinos: Alternativas para la conservación y uso sostenible en el Ecuador. INIAP, CIP, Quito, Lima, pp. 91–116.
Fanali, C., et al., 2011. Chemical characterization of sacha inchi (*Plukenetia volubilis* L.) oil. Journal of Agricultural and Food Chemistry 59, 13043–13049.
FAO, 2017a. FAOSTAT. Available at: http://www.fao.org/faostat/en/#data/QC. (Accessed 12 April 2019).
FAO, 2017b. The Future of Food and Agriculture - Trends and Challenges. FAO, Rome.

FAO, et al., 2018. The State of Food Security and Nutrition in the World 2018. Building Climate Resilence for Food Security and Nutrition. FAO, Rome.

FAO, Commission on Genetic Resources for Food and Agriculture Assessments, 2019. The State of the World's Biodiversity for Food an Agriculture. FAO, Rome.

FAO, PAR, Platform for Agrobiodiversity Research, 2010. Biodiversity for Food and Agriculture. FAO, Rome.

Fischer, G., 2000. Crecimiento y desarrollo. In: Flórez R., V.J., Fischer, G., Sora R., Á.D. (Eds.), Producción, poscosecha y exportación de la uchuva (*Physalis peruviana* L.). Universidad Nacional de Colombia, Santafé de Bogotá, pp. 9–26.

Fischer, G., Herrera, A.A.P.J., 2011. Cape gooseberry (*Physalis peruviana* L.). In: Yahia, E.M. (Ed.), Postharvest Biology and Technology of Tropical and Subtropical Fruits: Açai to Citrus. Woodhead Publishing, pp. 374–397.

Flores, H.E., et al., 2003. Andean root and tuber crops: underground rainbows. HortScience 38 (2), 161–167.

Fracassetti, D., Costa, C., Moulay, L., Tomás-Barberán, F.A., 2013. Ellagic acid derivatives, ellagitannins, proanthocyanidins and other phenolics, vitamin C and antioxidant capacity of two powder products from camu-camu fruit (*Myrciaria dubia*). Food Chemistry 139, 578–588.

Franco, L.A., et al., 2007. Actividad antinflamatoria de extractos y fracciones obtenidas de cálices de *Physalis peruviana* L. Biomedica 27, 110–115.

Franco, M.R.B., Shibamoto, T., 2000. Volatile composition of some Brazilian fruits: umbu-caja (*Spondias citherea*), camu-camu (*Myrciaria dubia*), Araça-boi (*Eugenia stipitata*) and Cupuaçu (*Theobroma grandiflorum*). Journal of Agricultural and Food Chemistry 48, 1263–1265.

Gallego Villa, D.Y., et al., 2014. Chemical and nutritional characterization of *Chenopodium pallidicaule* (cañihua) and *Chenopodium quinoa* (quinoa) seeds. Emirates Journal of Food and Agriculture 26 (7), 609–615.

Gamel, T.H., et al., 2004. Nutritional study of raw and popped seed proteins of *Amaranthus caudatus* L and *Amaranthus cruentus* L. Journal of the Science of Food and Agriculture 84, 1153–1158.

Genta, S.B., et al., 2010. Hypoglycemic activity of leaf organic extracts from *Smallanthus sonchifolius*: constituents of the most active fractions. Chemico-Biological Interactions 185, 143–152.

GPAFSN, Global Panel on Agriculture and Food Systems for Nutrition, 2016. Food Systems and Diets: Facing the Challenges of the 21st Century. Global Panel on Agriculture and Food Systems for Nutrition, London.

Graf, B.L., et al., 2015. Innovations in health value and functional food development of quinoa (*Chenopodium quinoa* Willd.). Comprehensive Reviews in Food Science and Food Safety 14, 431–445.

Gross, R., et al., 1989. Chemical composition and protein quality of some local Andean food sources. Food Chemistry 34, 25–34.

Gunstone, F.D., Pollard, M.R., Scrimgeour, C.M., Vedanayagam, H.S., 1977. Fatty acids. Part 50*. 13C nuclear magnetic resonance studies of olefinic fatty acids and esters. Chemistry and Physics of Lipids 18, 115–129.

Gutiérrez, L.-F., Rosada, L.-M., Jiménez, Á., 2011. Chemical composition of Sacha Inchi (*Plukenetia volubilis* L.) seeds and characteristics of their lipid fraction. Grasas y Aceites 62 (1), 76–83.

Hamaker, B.R., et al., 1992. Amino acid and fatty acid profiles of the Inca peanut (*Plukenetia volubilis*). Cereal Chemistry 69 (4), 461–463.

Heinrich, M., Dhanji, T., Casselman, I., 2011. Açai (*Euterpe oleracea* Mart.)-A Phytochemical and pharmacological assessment of the species' health claims. Phytochemistry Letters 4, 10–21.

Hind, N., 2010. Tropaeolum tuberosum. Curtis's Botanical Magazine 27 (3), 301–313.

HLPE, 2017. Nutrition and Food Systems. A Report by the High Level Panel of Experts on Food Security and Nutrition of the Committee on World Food Security. FAO, Rome.

INIA, 2019. Ministerio de Agricultura y Riego del Perú. Instituto Nacional de Innovación Agraria. Available at: http://www.inia.gob.pe/wp-content/uploads/investigacion/programa/sistProductivo/variedad/maiz-amilaceo. (Accessed 21 May 2019).

Jimenez, V.M., et al., 2014. Identification of phenolic compounds in soursop (*Annona muricata*) pulp by high-performance liquid chromatography with diode array and electrospray ionization mass spectrometric detection. Food Research International 65 (A), 42–46.

Koziol, M.J., 1992. Chemical composition and nutritional evaluation of quinoa (*Chenopodium quinoa* Willd.). Journal of Food Composition and Analysis 5, 35–68.

Krivankova, B., et al., 2012. Preliminary study of diversity of *Plukenetia volubilis* based on the morphological and genetic characteristics. Agricultura Tropica et Subtropica 45 (3), 140–146.

Lachman, J., Havrland, B., Fernández, E.C., Dudjak, J., 2004. Saccharides of yacon [*Smallanthus sonchifolius* (Poepp. et Endl.) H. Robinson] tubers and rhizomes and factors affecting their content. Plant Soil and Environment 50 (9), 383–390.

Lamothe, L.M., Srichuwong, S., Reuhs, B.L., Hamaker, B.R., 2015. Quinoa (*Chenopodium quinoa* W.) and amaranth (*Amaranthus caudatus* L.) provide dietary fibres high in pectic substances and xyloglucans. Food Chemistry 167, 490–496.

Lan, Y.H., et al., 2009. New cytotoxic withanolides from *Physalis peruviana*. Food Chemistry 116 (2), 462–469.

Lao, F., Sigurdson, G.T., Giusti, M.M., 2017. Health benefits of purple corn (*Zea mayz* L.) phenolic compounds. Comprehensive Reviews in Food Science and Food Safety 16, 234–246.

Li, C.-Y., et al., 2008. Corn husk as a potential source of anthocyanins. Journal of Agricultural and Food Chemistry 56, 11413–11416.

Li, X., et al., 2018. 1 introduction: setting the scene. In: Li, X., Siddique, K.H.M. (Eds.), Future Smart Food – Rediscovering Hidden Treasures of Neglected and Underutilized Species for Zero Hunger in Asia. FAO, Bangkok, pp. 15–32.

Li, G., Ammermann, U., Quirós, C.F., 2001. Glucosinolate contents in maca (*Lepidium peruvianum* Chacón) seeds, sprouts, mature plants and several derived commercial products. Economic Botany 55 (2), 255–262.

Lim, T.K., 2016. Tropaeolum tuberosum. In: Lim, T.K. (Ed.), Edible Medicinal and Non-medicinal Plants: Volume 12, Modified Stems, Roots, Bulbs. Springer International Publishing, pp. 94–102.

Lima Nascimento, A.K., et al., 2013. Antioxidant and antiproliferative activities of leaf extracts from *Plukenetia volubilis* linneo (Euphorbiaceae). Evidence-based Complementary and Alternative Medicine 1–10.

Lin, F., Hasegawa, M., Kodama, O., 2003. Purification and identification of antimicrobial sesquiterpene lactones from yacon (*Smallanthus sonchifolius*) leaves. Bioscience, Biotechnology, and Biochemistry 67 (10), 2154–2159.

MacLeod, A.J., Pieris, N.M., 1981. Volatile flavor components of soursop (*Annona muricata*). Journal of Agricultural and Food Chemistry 29, 488–490.

Meissner, H.O., et al., 2015. Peruvian maca: two scientific names Lepidium meyenii Walpers and Lepidium peruvianum chacon - are they phytochemically - synonymous? International Journal of Biomedical Science 11 (1), 1–15.

Mertz, C., et al., 2009. Phenolic compounds, carotenoids and antioxidant activity of three tropical fruits. Journal of Food Composition and Analysis 22, 381–387.

Mertz, C., Brat, P., Caris-Veyrat, C., Gunata, Z., 2010. Characterization and thermal lability of carotenoids and vitamin C of tamarillo fruit (*Solanum betaceum* Cav.). Food Chemistry 119, 653–659.

Moghadamtousi, S.Z., et al., 2015. *Annona muricata* (annonaceae): a review of its traditional uses, isolated acetogenins and biological activities. International Journal of Molecular Sciences 16, 15625–15658.

Muñoz, E.B., et al., 2018. Gamma-conglutin peptides from Andean lupin legume (*Lupinus mutabilis* Sweet) enhanced glucose uptake and reduced gluconeogenesis in vitro. Journal of Functional Foods 45, 339–347.

Nascimento, A.C., et al., 2014. Characterisation of nutrient profile of quinoa (*Chenopodium quinoa*), amaranth (*Amaranthus caudatus*), and purple corn (*Zea mays* L.) consumed in the North of Argentina: proximates, minerals and trace elements. Food Chemistry 148, 420–426.

Navruz-Varli, Semra, Sanlier, N., 2016. Nutritional and health benefits of quinoa (*Chenopodium quinoa* Willd.). Journal of Cereal Science 69, 371–376.

NRC, N.R.C., 1989. Lost Crops of the Incas: Little-Known Plants of the Andes with Promise for Worldwide Cultivation. The National Academies Press, Washington, DC.

Ohyama, T., et al., 1990. Composition of storage carbohydrate in tubers of yacon (*Polymnia sonchifolia*). Soil Science and Plant Nutrition 36 (1), 167–171.

Ortega, O.R., et al., 2006. Glucosinolate survey of cultivated and feral mashua (*Tropaeolum tubesorum* ruiz & pavon) in the cuzco region of Peru. Economic Botany 60 (3), 254–264.

Pacheco-Palencia, L., Duncan, C.E., Talcott, S.T., 2009. Phytochemical composition and thermal stability of two commercial açai species, *Euterpe oleracea* and *Euterpe precatoria*. Food Chemistry 115, 1199–1205.

Pedreschi, R., et al., 2003. Andean yacon root (*Smallanthus sonchifolius* poepp. Endl) fructooligosaccharides as a potential novel source of prebiotics. Journal of Agricultural and Food Chemistry 51, 5278–5284.

Pedreschi, R., Cisneros-Zevallos, L., 2007. Phenolic profiles of Andean purple corn (*Zea mays* L.). Food Chemistry 100, 956–963.

Peñarrieta, J.M., Alvarado, J.A., Åkesson, B., Bergenståhl, B., 2008. Total antioxidant capacity and content of flavonoids and other phenolic compounds in canihua (*Chenopodium pallidicaule*): an Andean pseudocereal. Molecular Nutrition and Food Research 52, 708–717.

Pinto, A.C.d.Q., et al., 2005. Taxonomy and botany. In: Williams, J.T., et al. (Eds.), Annona Species. International Centre for Underutilised Crops, Southampton, pp. 9–11.

Puente, L.A., Pinto-Muñoz, C.A., Castro, E.S., Cortés, M., 2011. *Physalis peruviana* Linnaeus, the multiple properties of a highly functional fruit: a review. Food Research International 44, 1733–1740.

Quiroga Ledezma, C.C., Ortiz Mérida, A.J., Escalera Vásquez, C.R., 2018. Evaluación de un Proceso Novedoso de Beneficiado en Seco del Grano de Cañahua (*Chenopodium pallidicuale* Aellen), Basado en la Aplicación de un Lecho Fluidizado Tipo Surtidor. Investigación and Desarrollo 18 (1), 17–34.

Quiroga, C., et al., 2015. Traditional processes and technological Innovations in quinoa harvesting, processing and industrialization. In: Bazile, D., Bertero, D., Nieto, C. (Eds.), State of the Art Report on Quinoa Around the World in 2013. FAO, CIRAD, Rome, pp. 218–249.

Quiroga, C.C., 2018. Quinua: el grano ancestral dorado de los Andes. In: Quiroz, E., Pradilla, H. (Eds.), El pasado del Futuro Alimentario: Los Alimentos Ancestrales Americanos. Instituto de Investigaciones Dr. José María Luis Mora, Ciudad de México, pp. 42–59.

Ramadan Hassanien, M.F., 2011. *Physalis peruviana*: a rich source of bioactive phytochemicals for functional foods and pharmaceuticals. Food Reviews International 27, 259–273.

Ramadan, M.F., 2011. Bioactive phytochemicals, nutritional value, and functional properties of cape gooseberry (*Physalis peruviana*): an overview. Food Research International 44, 1830–1836.

Ramos-Escudero, F., et al., 2012. Purple corn (*Zea mays* L.) phenolic compounds profile and its assessment as an agent against oxidative stress in isolated mouse organs. Journal of Medicinal Food 15 (2), 206–215.

Rastrelli, L., et al., 1995. Studies on the constituentes of *Amaranthus caudatus* (kiwicha) seeds. Isolation and characterization of seven new triterpene saponis. Journal of Agriculture and Food Chemisty 43, 904–909.

Rastrelli, L., De Simone, F., Schettino, O., Dini, A., 1996. Constituents of *Chenopodium pallidicaule* (cañihua) seeds: isolation and characterization of new triterpene saponins. Journal of Agriculture and Food Chemistry 3528–3533, 44.

Repo-Carrasco-Valencia, R., Hellström, J.K., Pihlava, J.-M., Mattila, P.H., 2010. Flavonoids and other phenolic compounds in Andean indigenous grains: quinoa (*Chenopodium quinoa*), kañiwa (*Chenopodium pallidicaule*) and kiwicha (*Amaranthus caudatus*). Food Chemistry 120, 128–133.

Rinderle, S.J., Goldstein, I.J., Matta, K.L., Ratcliffe, R.M., 1989. Isolation and characterization of amaranthin, a lectin present in the seeds of *Amaranthus caudatus*, that recognizes the T-(or cryptic T)-Antigen. Journal of Biological Chemistry 264 (27), 16123–16131.

Rodrigues, R.B., et al., 2001. An Amazonian fruit with a high potential as a natural source of vitamin C: the camu-camu (*Myrciaria dubia*). Fruits 56, 345–354.

Salinas Moreno, Y., et al., 2013. Caracterización Física y Composición Química de Razas de Maíz de Grano Azul/Morado en las regiones tropicales y subtropicales de Oaxaca. Revista Fitotecnia Mexicana 36 (1), 23–31.

Sathe, S.K., Kshirsagar, H.H., Sharma, G.M., 2012. Solubilization, fractionation, and electrophoretic characterization of Inca peanut (*Plukenetia volubilis* L.) proteins. Plant Foods for Human Nutrition 67, 247–255.

Schauss, A.G., et al., 2006. Phytochemical and nutrient composition of the freeze-dried amazonian palm berry, *Euterpe oleraceae* Mart. (Acai). Journal of Agricultural and Food Chemistry 54, 8598–8603.

Simonovska, B., et al., 2003. Investigation of phenolic acids in yacon (*Smallanthus sonchifolius*) leaves and tubers. Journal of Chromatography A 1016, 89–98.

Singh, N., Singh, S., Shevkani, K., 2019. Maize: composition, bioactive constituents, and unleavened bread. In: Preedy, V., Watson, R., Patel, V. (Eds.), Flour and Breads and Their Fortification in Health and Disease Prevention. Elservier Inc., San Diego, pp. 89–100.

Takenaka, M., et al., 2003. Caffeic acid derivatives in the roots of yacon (*Smallanthus sonchifolius*). Journal of Agricultural and Food Chemistry 51, 793–796.

Tang, Y., et al., 2015. Characterisation of fatty acid, carotenoid, tocopherol/tocotrienol compositions and antioxidant activities in seeds of three *Chenopodium quinoa* Willd. genotypes. Food Chemistry 174, 502–508.

Tang, Y., Tsao, R., 2017. Phytochemicals in quinoa and amaranth grains and their antioxidant, anti-inflammatory, and potential health beneficial effects: a review. Molecular Nutrition and Food Research 61 (7), 1–16.

Tapia, M.E., Fries, A.M., 2007. Guía de campo de los cultivos andinos, first ed. FAO; ANPE, Lima.

Ulbricht, C., et al., 2012. An evidence-based systematic review of Açaí (Euterpe oleracea) by the Natural Standard Research Collaboration. Journal of Dietary Supplements 2 (9), 128–147.

UN, 2019. Sustainable Development Goals. Available at: https://www.un.org/sustainabledevelopment/hunger/. (Accessed 3 April 2019).

Valentova, K., et al., 2003. Antioxidant activity of extracts from the leaves of *Smallanthus sonchifolius*. European Journal of Nutrition 42, 61–66.

Valerio Jr., L.G., Gonzales, G.F., 2005. Toxicological aspects of the South American herbs cat's claw (*Uncaria tomentosa*) and maca (*Lepidium meyenii*). Toxicological Reviews 24 (1), 11–35.

Vijayalakshmi, G., Rajeswari V, D., 2012. Phytochemical and pharmacological properties of *Annona muricata*: a review. International Journal of Pharmacy and Pharmaceutical Sciences 4 (2), 3–6.

Villacrés, E., et al., 2010. Evaluación del Rendimiento, Características Fisico-Químicas y Nutraceúticas del Aceite de Chocho (*Lupinus mutabilis* sweet). Revista Tecnológica ESPOL - RTE 23, 57–62.

Wang, S., Zhu, F., Kakuda, Y., 2018. Sacha inchi (*Plukenetia volubilis* L.): nutritional composition, biological activity, and uses. Food Chemistry 265, 316–328.

White, P.L., et al., 1955. Nutritive values of crops. Nutrient content and protein quality of quinua and cañihua, edible seed products of the Andes mountains. Agricultural and Food Chemistry 3 (6), 531–534.

WHO, 2018a. Malnutrition, Key Facts. Available at: https://www.who.int/news-room/fact-sheets/detail/malnutrition. (Accessed 11 April 2019).

WHO, 2018b. Noncommunicable Diseases, Key Factors. Available at: https://www.who.int/news-room/fact-sheets/detail/noncommunicable-diseases. (Accessed 11 April 2019).

Wu, S.-J., et al., 2009. Supercritical carbon dioxide extract of *Physalis peruviana* induced cell cycle arrest and apoptosis in human lung cancer H661 cells. Food and Chemical Toxicology 47, 1132–1138.

Yan, X., et al., 1999. Extraction and identification of antioxidants in the roots of yacon (*Smallantus sonchifolius*). Journal of Agricultural and Food Chemistry 47, 4711–4713.

Yazawa, K., et al., 2011. Anti-Inflammatory effects of seeds of the tropical fruit camu-camu (*Myrciaria dubia*). Journal of Nutritional Science and Vitaminology 57, 104–107.

Zanatta, C.F., Mercadante, A.Z., 2007. Carotenoid composition from the Brazilian tropical fruit camu-camu (*Myrciaria dubia*). Food Chemistry 101, 1526–1532.

Zapata, S.M., Dufour, J.-P., 1993. Camu-camu *Myrciaria dubia* (HBK) McVaugh: chemical composition of fruit. Journal of the Science of Food and Agriculture 61, 349–351.

CHAPTER

Fermenting locally to improve soil fertility: global lessons to the future of food sovereignty

2

Salvador Roig-Coll, MSc, Andriana Kechagia, MSc
Terrapi World S.L., Ibiza, Spain

1. By-products and fermentation

Agro-industrial residues encompass a wide range of organic material including agriculture, animal, and industrial by-products (Fig. 2.1). It is estimated that approximately one-third of the food produced for human consumption is wasted yearly (Gustavsson et al., 2011). Only in Europe, around 88 million tons of edible food are wasted per year, with a cost of €143 billion (Fusions, 2016).

These by-products may be used as raw material for the production of fertilizers, pesticides, biogas, biofuel, mushroom, and tempeh, while they are considered to be a source of bioactive compounds (Sadh et al., 2018). The production is based on

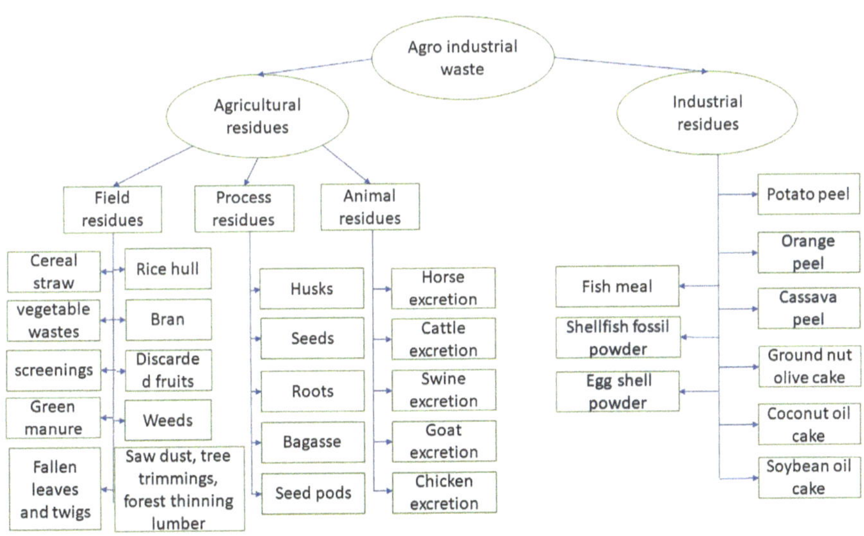

FIGURE 2.1

Examples of agro-industrial residues worldwide.

Sustainability of the Food System. https://doi.org/10.1016/B978-0-12-818293-2.00002-1

fermentation processes that make use of microorganisms. The microorganisms are fungi, bacteria, and/or yeasts that are favored by the high-water absorption capacity of the residues, while they exhibited a good growth rate (Orzua et al., 2009).

The term fermentation is used to describe an anaerobic process where only certain microorganisms can act. Those differ to the organisms that act in the composting processes since they are active in the absence of oxygen. However, there are organisms such as *Bacillus thuringiensis*, which are facultative anaerobic and have the capacity of acting in both composting and fermenting processes. During fermentation, organic acids, e.g., lactic acid, butyric acid, and acetic acid (vinegar), and biologically "active" compounds, e.g., antibiotics such as streptomycin, are produced. Examples of microorganisms cultured in agricultural residues are *B. thuringiensis* (Zhang et al., 2013) and *Trichoderma harzianum* (Lihua et al., 2011). As a result, bioorganic fertilizers are produced that control crop pests, mosquitos, or fusarium wilt with high efficacy (Lihua et al., 2011). There is a dual impact of such products in socioeconomics by reducing the inputs cost and by reducing pollution and promoting the sustainability concept.

A combination of diverse derived residues goes through a fermentation process, usually a solid-state fermentation (SSF). SSF occurs when microorganisms grow on moistened solid substrate. The level of moisture during SSF is adequate to support microbial growth and metabolism (Rahardjo et al., 2006). Traditionally, SSF has been used for centuries mainly for producing fermented foods (Campbell-Platt, 1994). In Asia, SSF is part of the culture, whereas fermented foods are a source of healthy nutrition and a source of income to communities. Fermentation served as a typical method to increase food production and to prolong food shelf life. Examples of fermented foods are the tempe in Indonesia, kimchi in Korea, appam in India, and gundruk in Nepal (FAO, 2016). Furthermore, fermented food has been a strategy for emergency food storage to cope with unpredicted phenomena in Polynesia (Ragone, 1991; Lindberg and Mossing, 1996).

2. Soil regeneration applying fermentation techniques

2.1 Korea

In Korean traditionally natural farming, farmers collect and culture whole communities of microorganisms without specificity. Indigenous microorganisms are added in the containers of agrowaste, where they ferment and convert into biofertilizers (Cho et al., 2010). Then, the fertilizers are applied to the plants through the foliar part nourishing the plant effectively for every growth stage. When the biofertilizer containing indigenous organisms was applied on leafy vegetables such as leek, beet, curly kale, and burdock, they exhibited a higher growth (Wang et al., 2013). Also, the biofertilizers showed to have a better capacity than conventional treatments for weed control (Wang et al., 2013). Consequently, farmers maximize the use of farm inputs, recycling farm waste while fostering and building up soil fertility.

Korean farmers achieve sustainability by promoting microorganisms diversity, increasing the production and well-nourished plants, controlling weeds and pests, recycling any agrowaste, and protecting the environment.

2.2 Yunnan, China

In ecological recycling production system, pig wastes are being fermented with microorganism inoculums in their liquid form and are directly used as liquid fertilizer in banana production. Yuanjia Agricultural Development Co., Ltd. have created an innovation based on a "Livestock (pig)—Waste—Banana" ecological recycling production system in Yuanyang, Yunnan. 5000 tons of liquid manure is used in banana drip irrigation system. In this way, resistance against fusarium genotypes was observed. The fermented liquid manure contained beneficial microorganisms such as *Bacillus*, which were found to be the most abundant bacterial group in the disease-suppressing soils in the banana rhizosphere. The plant health status was correlated to the composition of the soil microbial community (Xue et al., 2015). It seems that microbial treatments in the soil amendment enhance and boost crop yields (Wang et al., 2018). As a result, soil borne diseases were suppressed whereas soil fertility, soil organic matter content and yield and quality of banana increased.

2.3 India

Indian farmers are producing the Jeevamrit (biozyme). Jeevamrit is made by available farm wastes such as cattle dung and cattle urine along with other ingredients such as jaggery, pulse flour, and water. It is a fermented input in crop production at low cost. The philosophy behind the application is the supplementation of essential plant nutrients and improvement of the soil health in economic and eco-friendly manner (reddy, 2008). It was found that biozyme consists of significant amounts of essential macro- and micronutrients, vitamins, amino acids, and plant growth—promoting substances such as indole acetic acid and gibberellic acid. Applying the biozyme resulted in higher growth and yield in the plant as well as an increase in the microbial population of the soil.

2.4 Japan

Bokashi is produced in Japan using crop residues, animal wastes, organic material, food processing residues, and mineral resources of natural origin. The application of bokashi in the soil affects plant nutrition directly and indirectly: directly through the availability of the nutrients in the rhizosphere and indirectly favoring microorganism activities in the rhizosphere. Microorganisms involved in fermentation include *Rhyzopus*, Koji mold (*Aspergillus oryzae*), *Bacillus natto, Bacillus subtilis*, lactic acid bacilli, yeast, and *Actinomycetes*, which generate organic acids, amino acids, vitamins, plant hormones, enzymes, and minerals. As a result, the crop exhibits a vigorous and healthy growth. They reduce activities of pathogenic microorganisms, fungi, bacteria, and nematodes (Wakui, 2009).

"Tsuchi dukuri" refers to soil enrichment practices using available resources locally. Farm, animal, and food wastes are incorporated into the soil, reducing the cost of fertilization inputs (Wakui, 2009).

2.5 Balearic islands, Spain

In Europe, such fermentation techniques for farm management exhibit a rather innovative approach since there is no tradition. In 2018, S.A.T. Agronatura (Mallorca Island) in collaboration with Terrapi World S.L. have created an innovative fermented solution enriched with local microorganisms to treat cereal seeds before planting. The objective was to evaluate the effect of local bacteria in improving soil conditions and crop growth for rainfed cereal production. The results showed that there was a significant effect of the bacteria inoculum on the seeds in the rhizosphere microbial ecosystem. The root development during the initial crop growth was faster than in the no-inoculated seeds, suggesting that microbial inoculum did stimulate root growth. Furthermore, the chemical analysis showed that the level of nutrients was higher in the treated plants, indicating a positive effect of the inoculum in nutrient and water uptake and thus an enhanced crop nutrition.

Another case is the treatment of sewage sludge to improve the soil organic matter. In 2017, Terrapi World S.L. in collaboration with the Municipality of Santa Eularia des Riu (Ibiza Island) set up a pilot to improve the biological quality of sewage sludge. When sewage sludge was treated with selected local bacteria under anaerobic digestion, the population of the pathogen *Clostridium perfringens* was reduced to undetectable counts (Fig. 2.2A). That result may happen because of

FIGURE 2.2

(A) Sewage sludge treated with selected local bacteria under anaerobic digestion. (B) Plant growth of barley grown on soil treated with the fermented sludge (i) and without treatment (ii).

Credits: Andriana Kechagia.

the medium acidification. Lactic fermentation creates lactic acid, which preacts as prebiotics for further development of acidophilic bacterial colonies in the sludge and in the soil (Moreno Caso & Mormeneo Bernat, 2007). Consequently, the sewage sludge was considered as a safe material for soil amendment rich in organic matter.

Fig. 2.2B shows that this soil amendment favored the growth of more robust barley plants in comparison with an untreated soil. The high content of organic matter in combination with the microbial inoculum stimulated root growth and biomass.

3. Lessons for food sovereignty

All these techniques that have been used locally through the whole world represent the same concept of sustainability respecting the diversity of ecosystems. The pros of such methods are as follows:

- Local communities have developed agricultural practices that encompass the three properties of the soil: chemical, biological, and physical, simultaneously respecting the natural diversity and the systems dynamic. Each one of the properties named before is equally important standing alone and is combined to sustain crop productivity, environmental quality, and soil microorganisms (Jackson et al., 2003). There is a fundamental difference with the commercial strategy that focuses only on the chemical properties of the soil quality, minimizing the importance of the biological and physical ones.
- The recycling of residues produced locally.
- The culture of diverse microorganisms and their enzymes: According to Signh, the loss of microbial diversity is a major reason for deterioration of soil fertility and agricultural productivity (Signh et al., 2011).
- The plant and/or animal source used as a soil amendment is diverse, resulting in a rich nutritious input consisting of a wide range of micro- and macronutrients rather than a typical three-element fertilization.
- Soil fertility is enhanced.
- The fertilizer input is of low cost.
- The suppressed soil results in a better control of pests, avoiding extra costs and further environment pollution.
- The N leaching into the environment decreases. It is shown that soil NO_3−N accumulation decreases when N is applied in combination with P and/or manure (Tong et al., 1997; Hao et al., 2005). Instead, root growth is enhanced and resulted in higher nutrient uptake and crop yield (Wen et al., 2016).

References

Campbell-Platt, G., 1994. Fermented foods — a world perspective. Food Research International 27 (3), 253–257. ISSN 0963-9969.

Cho, H.-K., Cho, J.-Y., 2010. Cho Han-Kyu Natural Farming. Cho Global Natural Farming, Seoul, Korea.

FAO, 2016. In: Proceedings of the FAO International Symposium on the Role of Agricultural Biotechnologies in Sustainable Food Productions Systems and Nutrition. ISBN 978-92-5-109330-6.

Fusions, Stenmarck, A., Jensen, C., Quested, T., Moates, G., 2016. Estimates of European Food Waste Levels. EU FP7 Fusions.

Gustavsson, J., Cederberg, C., Sonesson, U., Emanuelsson, A., 2011. The Methodology of the FAO Study: "Global Food Losses and Food Waste-Extent, Causes and Prevention" – FAO. SIK, Göteborg.

Hao, M.D., Fan, J., Wei, X.R., Peng, L.F., Lai, L., 2005. Effects of fertilization on soil fertility and wheat yield of dryland wheat in the Loess Plateau. Pedosphere 15, 189–195.

Jackson, L.E., Calderon, F.J., Steenwerth, K.L., Scow, K.M., Rolston, D.E., 2003. Responses of soil microbial processes and community structure to tillage events and implications for soil quality. Geoderma 114, 305–317.

Lihua, C., Xingming, Y., Waseem, R., Jia, L., Fengge, Z., Qirong, S., 2011. Solid-state fermentation of agro-industrial wastes to produce bioorganic fertilizer for the biocontrol of Fusarium wilt of cucumber in continuously cropped soil. Bioresource Technology 102 (4), 3900–3910. ISSN 0960-8524.

Lindberg, P., Mossing, A., 1996. The Effects of Cyclones on Agriculture in Western Samoa. Department of Physical Geography, Umeå University, Sweden.

Moreno Casco, J., Mormeneo Bernat, S., 2007. Microbiología y bioquímica del proceso de compostaje. In: Compostaje, pp. 111–140.

Orzua, M.C., Mussattob, S.I., Contreras, E.J.C., Rodrigueza, R., Garzaa, H.T.A., Aguilara, C.N., 2009. Exploitation of agro-industrial wastes as immobilization carrier for solid-state fermentation. Industrial Crops and Products 30, 24–27.

Ragone, D., 1991. Ethnobotany of breadfruit in Polynesia. In: Cox, P.A., Banack, S.A. (Eds.), Islands, Plants and Polynesians. An Introduction to Polynesian Ethnobotany. Dioscorides Press, Portland, OR, pp. 203–220.

Rahardjo, Y.S.P., Tramper, J., Rinzema, A., 2006. Modeling conversion and transport phenomena in solid-state fermentation: a review and perspectives. Biotechnology Advances 24, 161–179.

Sadh, P., Duhan, S., Duhan, J., 2018. Agro-industrial wastes and their utilization using solid state fermentation: a review. Bioresources and Bioprocessing 5. https://doi.org/10.1186/s40643-017-0187-z.

Singh, J.S., Pandey, V.C., Singh, D.P., 2011. Efficient soil microorganisms: a new dimension for sustainable agriculture and environmental development. Agriculture, Ecosystems and Environment 140, 339–353.

Tong, Y.N., Emteryd, O., Lu, D.Q., Grip, H., 1997. Effect of organic manure and chemical fertilizer on nitrogen uptake and nitrate leaching in a Eum-orthic anthrosols profile. Nutrient Cycling in Agroecosystems 48, 225–229.

Wakui, Y., 2009. Organic Farming Technology in Japan. Pilot Project for Better Farm Income by Organic-Based Vegetable Production.

Wang, K.-H., DuPonte, M., Chang, K., 2013. Use of Korean Natural Farming for Vegetable Crop Production in Hawai'i. Hānai'Ai/The Food Provider.

Wang, J., Li, Q., Xu, S., Zhao, W., Lei, Y., Song, C., Huang, Z., 2018. Traits-Based integration of multi-species inoculants facilitates shifts of indigenous soil bacterial community. Frontiers in Microbiology 9, 1692. https://doi.org/10.3389/fmicb.2018.01692.

Wen, Z., Shen, J., Martin, B., Li, H., Zhao, B., Yuan, H., 2016. Combined applications of nitrogen and phosphorus fertilizers with manure increase maize yield and nutrient uptake via stimulating root growth in a long-term experiment. Pedosphere 26 (1), 62–73.

Xue, C., Penton, C.R., Shen, Z., Zhang, R., Huang, Q., Li, R., Ruan, Y., Shen, Q., 2015. Manipulating the banana rhizosphere microbiome for biological control of Panama disease. Scientific Reports 5, 11124. https://doi.org/10.1038/srep11124.

Zhang, W., Qiu, L., Gong, A., Cao, Y., Wang, B., 2013. Solid-state fermentation of kitchen waste for production of *Bacillus thuringiensis*-based bio-pesticide. BioResources 8. https://doi.org/10.15376/biores.8.1.1124-1135.

CHAPTER

Climate change and agriculture: carbon footprint estimation for agricultural products and labeling for emissions mitigation

3

V.D. Litskas[1,2], D.P. Platis[1], C.D. Anagnostopoulos[1], A.C. Tsaboula[1], G.C. Menexes[3], K.L. Kalburtji[3], M.C. Stavrinides[2], A.P. Mamolos[1]

[1]*School of Agriculture, Laboratory of Ecology and Environmental Protection, Aristotle University of Thessaloniki, Thessaloniki, Greece;* [2]*Department of Agricultural Sciences, Biotechnology and Food Science, Cyprus University of Technology, Limassol, Cyprus;* [3]*School of Agriculture, Laboratory of Agronomy, Aristotle University of Thessaloniki, Thessaloniki, Greece*

Abbreviations

CC	Climate change
CF	Carbon footprint
CH_4	Methane
CO_2	Carbon dioxide
GHG(s)	Greenhouse gase(s)
GHGE(s)	Greenhouse gas emission(s)
IPCC	Intergovernmental Panel on Climate Change
LC	Life cycle
LCA	Life Cycle Assessment
N_2O	Nitrous oxide
PEF	Product Environmental Footprint
PEFCR	Product Environmental Footprint Category Rules

1. Climate change and emissions from agriculture

1.1 Climate change

Climate change (CC) is expected to substantially impact agriculture and food production (Hoegh-Guldberg et al., 2018), a trillion US dollars industry (Fig. 3.1). In semiarid areas of the world, such as the Mediterranean basin, extreme drought and

Sustainability of the Food System. https://doi.org/10.1016/B978-0-12-818293-2.00003-3

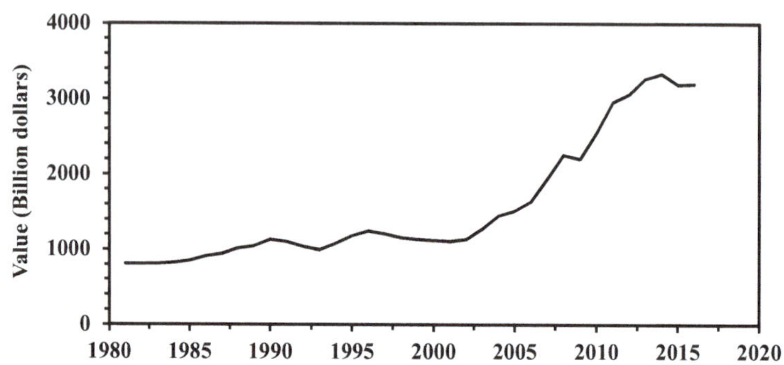

FIGURE 3.1

Agriculture, forestry, and fishing value (current US$).

Developed after using data from: World Bank, (2019). Agriculture, forestry, and fishing, value added (current US$). https://data.worldbank.org/indicator/NV.AGR.TOTL.CD?end=2017&start=1980.

heat have already impacted the production of crops, especially of high-income crops, such as fruits and vegetables (Iglesias and Garrote, 2015). The significant decrease in precipitation in countries of the Mediterranean basin can result to severe crop losses (Carnicer et al., 2011; Ponti et al., 2014). Variations in crop yields are expected to increase from year to year due to extreme weather events and other factors such as pests and diseases (Litskas et al., 2019a; Savary et al., 2019). Climate change has already impacted the length of the growing season, especially at higher latitudes (Liu et al., 2009; Teixeira et al., 2013). In North Europe, flowering and harvest dates for cereals (e.g., wheat, burley) occur earlier compared with previous decades (Rötter et al., 2012). On the other hand, in these areas, the increased frequency of flood incidents and heat waves, during the summer, might severely impact crops (Rojas et al., 2013).

Greenhouse gas emissions (GHGEs) from several sectors of the economy are the main drivers of CC. The 21st session of the conference of the parties in Paris (United Nations, The Paris Agreement) stressed the importance of reducing GHGEs to limit global warming at well below 2 °C relative to preindustrial levels (Hoegh-Guldberg et al., 2018). Reduction of GHGEs can be achieved by decreasing the use of natural resources (e.g., fuel, water) and by altering farming practices (Litskas et al., 2017; Michos et al., 2018). However, stabilizing the temperature increase to below 2 °C relative to preindustrial levels will require an urgent and fundamental departure from business as usual (Hoegh-Guldberg et al., 2018). The IPCC fifth assessment report (Smith et al., 2013) highlights that there are available ways to limit CC and its risks, with many solutions that allow for continued economic and human development. Recent policies on agriculture worldwide tend to reduce inputs (e.g., fertilizers, water, fuels) while maintaining high yields (ECF, 2010; Abeliotis et al., 2013).

1.2 Greenhouse gas emissions from agriculture and animal husbandry

At a global scale, agriculture accounts for 20% of the anthropogenic GHGEs (Hoegh-Guldberg et al., 2018). In the EU, agriculture emits 470.6 Mt of CO_2–equivalent per year, which corresponds to 10% of the total GHGEs (Eurostat, 2017). Furthermore, food is processed, stored, packaged, transported, prepared, and served to reach the consumers. All these stages demand energy and materials, and consequently, greenhouse gases are released into the atmosphere (EEA, 2015).

Animal husbandry is also considered as an activity contributing to CC, because of the release of methane (CH_4) and nitrous oxide (N_2O) directly from animals and indirectly from manure degradation (Chadwick et al., 2011; Herrero et al., 2013, 2016; Hou et al., 2015). The policies implemented in the EU (1990–today) resulted in a decrease in livestock numbers and the respective emissions (Herrero et al., 2016). However, policies and management practices for the reduction of GHGEs from livestock are not applied throughout the globe (CGIAR, 2019). Greenhouse gas emissions from livestock production increased in the developing countries during the past decades (Table 3.1.) (CGIAR, 2019; Herrero et al., 2016). As a result, emissions from enteric fermentation increased 11% in this period (1990–2012) and accounted for 39% of the sector's total GHG outputs in 2012 (CGIAR, 2019).

Table 3.1 Non-CO_2 emissions (current and projected) from the global agricultural sector.

Type of emission	Total emissions 1990 ($MtCO_2e$)	Total emissions 2012 ($MtCO_2e$)	Total emissions 2030 (projection) ($MtCO_2e$)
	(Million tons of CO_2 equivalent)		
Agricultural soils (N_2O)	1614 (35.4%)	2114 (39.3%)	2195 (38.1%)
Enteric fermentation (CH_4)	1869 (41%)	2080 (38.7%)	2365 (41.1%)
Rice cultivation (CH_4)	466 (10.2%)	522 (9.7%)	502 (8.7%)
Manure management (CH_4, N_2O)	335 (7.3%)	364 (6.8%)	413 (7.2%)
Other emissions (CH_4, N_2O)	277 (6.1%)	301 (5.6%)	281 (4.9%)
Total non-CO_2 emissions	4561 (100%)	5381 (100%)	5756 (100%)

CGIAR, 2019. Direct Agricultural Emissions. Available at: https://ccafs.cgiar.org/bigfacts/#theme=food-emissions&subtheme=direct-agriculture.

2. Using life cycle assessment and carbon footprint for emissions estimation

2.1 Life cycle assessment for agricultural products

Life cycle assessment (LCA) is a method for measuring the environmental impact of products (Pergola et al., 2017). Life cycle assessment is used to collect all the relevant data to estimate the carbon footprint (CF) for a product (Michos et al., 2018). The main benefit of the method is the assessment of GHGEs related to an individual production system. In this way, production improvements and different system comparisons could be implemented (Hillier et al., 2011, 2012). This kind of analysis of farming systems could lead to the best management practices (Litskas et al., 2013; Michos et al., 2018).

A popular LCA approach bases the estimation of CF on ISO standards (ISO, 2006a, 2006b, 2013). The successful implementation of the CF of an agricultural product needs the collection of detailed information regarding inputs and outputs of production and their individual use (Litskas et al., 2017).

2.2 Carbon footprint

Carbonfootprint is the sum of GHGs (CO_2, CH_4, N_2O) emitted during production, use, and final disposal of a product, expressed per kg of product (Clune et al., 2017; Michos et al., 2018). The GHGEs from fuels, fertilizers, pesticides, irrigation, and soil management (including land use changes) are calculated by the respective coefficients of CO_2-equivalent (CO_2-eq) according to protocols such as those provided by IPCC (1997, 2006, 2014) and EMEP/EEA (2009). Additionally, factors such as market demand, distance to market (e.g., emissions due to transportation), and climatic conditions (e.g., fewer optimum conditions, more inputs for agricultural production) could affect the CF. The CO_2-eq is used as a reference unit for the CF (synonymous with global warming potential) (Clune et al., 2017).

Measuring the CF can pinpoint emission "hotspots" and lead to reductions of GHGEs while improving production efficiency (Clune et al., 2017; Michos et al., 2018). Life cycle assessment for CF determination results in the identification of input overuse, which typically leads to increased costs. The management of the cultivation practices during the whole production process, the means of transport, and the use of machinery and fertilizers are among the factors that could optimize the management of inputs and decrease the CF of an agricultural product (Kehagias et al., 2015). By optimizing the inputs for production, the product cost decreases, enhancing in this way the desirable competitiveness on the target markets (Hartikainen et al., 2014). Therefore, the CF is an effective management evaluation tool for environmentally friendly agricultural production. Consumers, traders, and food industries could contribute to more balanced management of inputs.

3. Carbon footprint of agricultural products and management practices for its reduction

3.1 Carbon footprint of different agricultural products

Clune et al. (2017) implemented a systematic literature review of GHGEs for different food products from LCA studies. This was the first effort to gather all the data related to the CF of agricultural products. The summary of results for the CF values recorded in this work for key crops is presented in Table 3.2, whereas CF values for individual food products are presented in Fig. 3.2. In the work of Clune et al. (2017), most of the CF data come from developed countries.

In general, vegetables grown outdoors (e.g., onion, carrot) usually have low CF, whereas intensively managed fruits (e.g., peach, cherry) have a higher CF, due to higher inputs. The highest CF is observed for vegetables grown in greenhouses and rice. High CF was also estimated in several studies for fish, meat, and dairy products, whereas the highest CF values were found for meat from ruminants (Clune et al., 2017; Litskas et al., 2017, 2019b).

3.2 Carbon footprint "hotspots"

The CF "hotspots" vary for different crops (Litskas et al., 2019b). One of the most important CF "hotspots" for intensive agriculture is fertilizer use. Fertilizer production consumes high amounts of energy, resulting in substantial GHGEs (Hillier et al., 2012). In addition, fertilizer application leads to N_2O emissions that derive from the transformations occurring in the nitrogen cycle in soils (Chadwick et al., 2011; Herrero et al., 2013; Hou et al., 2015) (Fig. 3.3).

Field energy use is another important CF "hotspot" (Fig. 3.3). Energy production and consumption (e.g., for tillage, irrigation, transportation) contributes substantially to GHGEs (Kaltsas et al., 2007; Kavargiris et al., 2009; Litskas et al., 2017; Michos et al., 2018). The assessment of energy use in agriculture is crucial, since

Table 3.2 Individual carbon footprint (CF) values (kg CO_2-eq/kg product) for different products, from low to high (Clune et al., 2017).

Product	Median CF (kg CO_2-eq/kg product)
Vegetables (all field-grown vegetable)	0.37
Fruits (all field-grown fruit)	0.42
Cereals	0.50
Legumes	0.51
Passive (unheated) greenhouse fruit and vegetable	1.10
Tree nuts combined	1.20
Heated greenhouse fruits and vegetables	2.13
Rice	2.55

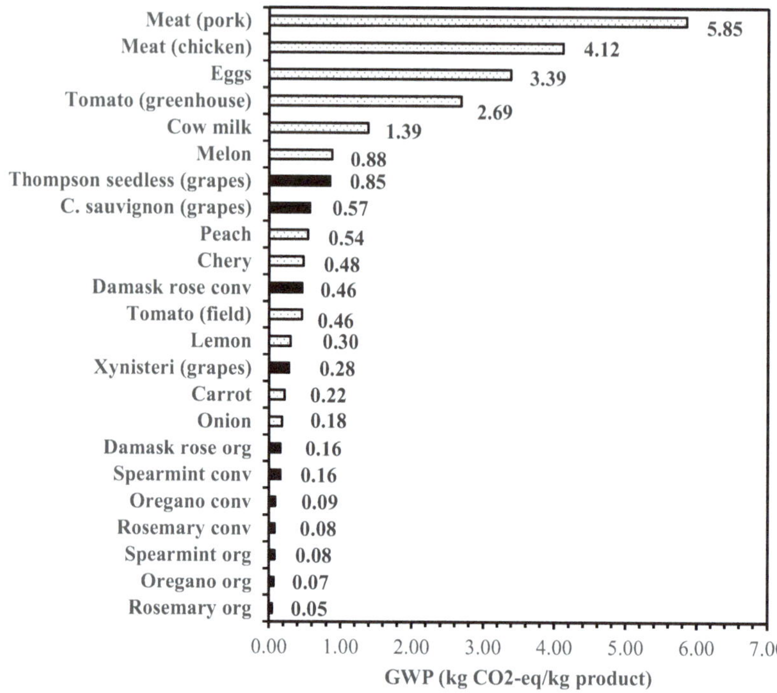

FIGURE 3.2

Indicative carbon footprint (CF) values for different organic (org.) and conventional (conv.) food and plant products.

Data to construct the graph were obtained from the tables presented in Clune, S., Crossin, E., Verghese, K., 2017. Systematic review of greenhouse gas emissions for different fresh food categories. Journal of Cleaner Production 140, 766–783. https://doi.org/10.1016/j.jclepro.2016.04.082; (whitebars) and Litskas, V.D., Irakleous, T., Tzortzakis, N. & Stavrinides, M.C. (2017). Determining the carbon footprint of indigenous and introduced grape varieties through life cycle assessment using the island of Cyprus as a case study. Journal of Cleaner Production 156, 418–425. https://doi.org/10.1016/j.jclepro.2017.04.057; Litskas D.V., Chrysargyris, A., Stavrinides, C.M., Tzortzakis, N. (2019b). Water-energy-food nexus: a case study on medicinal and aromatic plants. Journal of Cleaner Production. 233 1; (blackbars).

it is mainly in a nonrenewable form (Kavargiris et al., 2009; Michos et al., 2017). The agricultural sector accounts for 5% of the total energy consumption in the world (IPCC, 2014). Energy use can be evaluated via an energy analysis of agricultural systems (Litskas et al., 2019b), an approach based on the conversion of all the production factors and crop products into energy units (Litskas et al., 2011, 2013; Michos et al., 2012, 2017, 2018).

Other important CF "hotspots" include residue management in the field, pesticide applications and transportation of the products to the market (Litskas et al., 2017, 2019b).

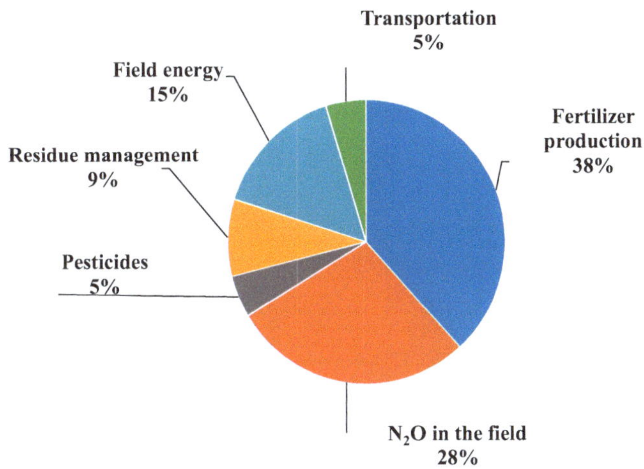

FIGURE 3.3

Carbon footprint "hotspots" determined for introduced and indigenous vine varieties in Cyprus.

Data to construct the graph were obtained from Litskas, V.D., Irakleous, T., Tzortzakis, N., Stavrinides, M.C. (2017). Determining the carbon footprint of indigenous and introduced grape varieties through life cycle assessment using the island of Cyprus as a case study. Journal of Cleaner Production 156, 418–425. https://doi.org/10.1016/j.jclepro.2017.04.057.

Besides management practices at farm scale, land use change and deforestation also contribute to substantial emissions. Almost 50% of the Agriculture, Forestry, and Other Land Use (AFOLU) emissions derive from land-use change and deforestation (IPCC, 2014).

3.3 Carbon footprint reduction

Intensive agriculture leads to higher input use and increased GHGEs compared with low input agriculture, such as organic farming (Clune et al., 2017). Several authors proposed that organic and integrated farming contribute less to global warming than conventional farming (Flessa et al., 2002; Hoeppner et al., 2006; Olesen et al., 2006; Kaltsas et al., 2007; Michos et al., 2012, 2018; Litskas et al., 2017). A recent study (Litskas et al., 2019b) showed that organic farming led to lower CF for spearmint and damask rose than conventional production. However, there were no significant differences in the CF between organically and conventionally produced oregano and rosemary.

Use of local varieties that are heat and drought resistant could be valuable toward emissions mitigation (Sadras and Lawson, 2013; Litskas et al., 2017). Adapted varieties are by definition well suited to the local climate and soil conditions and consequently require less inputs resulting to reduced CF. As a representative example, in

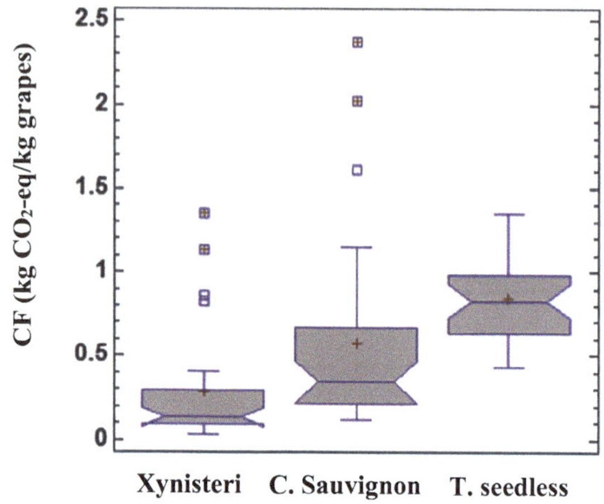

FIGURE 3.4

Carbon footprint (CF) values for indigenous and introduced grape varieties cultivated in Cyprus. Xynisteri (local white), C. Sauvignon (Cabernet Sauvignon—introduced), T. seedless (Thompson seedless—introduced). The whiskers extend at 1.5 times the interquartile range, and the outliers are also plotted (as dots). The box gives the 25th and 75th percentile, and the line in it is the median value. The cross is the mean value, and the points (if present) above/below the whiskers are outliers.

Data from Litskas, V.D., Irakleous, T., Tzortzakis, N. & Stavrinides, M.C. (2017). Determining the carbon footprint of indigenous and introduced grape varieties through life cycle assessment using the island of Cyprus as a case study. Journal of Cleaner Production 156, 418–425. https://doi.org/10.1016/j.jclepro.2017.04.057 where used for the construction of this figure.

Fig. 3.4, the CF comparison is presented for indigenous and introduced grape varieties in Cyprus using data from Litskas et al. (2017).

Farming practices can change to minimize GHGEs, e.g., rotating crops to take advantage of the soil water regime (Zhou et al., 2011; Mohammad et al., 2012). Other options are to adjust sowing to the changed rainfall patterns and use crop varieties better suited to the new conditions (Asseng et al., 2019) including native varieties. Mitigation of GHGEs could be also achieved by managing soils to store C, rather than emitting CH_4 and N_2O (Litskas et al., 2017). Organic fertilizers could be used for this purpose (Hillier et al., 2012; Litskas et al., 2017). Soil emissions are significant GHGEs sources (Albaladejo et al., 2013), despite the fact that more efficient production and use of fertilizers in combination with better manure management could reduce the GHGEs. For example, more efficient fertilizer production and management reduced the EU's emissions from agriculture by 24% between 1990 and 2012 (EEA, 2015).

Current policies focus on minimizing deforestation to avert CO_2 emissions from land use change (Popp et al., 2014) or on enhancing more efficient agronomic practices for reduction of GHGEs in agricultural production processes (Martin et al., 2010; Smith et al., 2013). Besides these policies, changing dietary habits and promoting agricultural products with reduced CF could result in lower GHGEs (Stehfest et al., 2009; Popp et al., 2010). According to IPCC (2014), among the key mitigation activities related to the agricultural sector are the following:

- Reduction of CH_4 or N_2O emissions from crop production.
- Reductions of direct (e.g., use of machinery, use of fertilizers) or indirect GHGEs (e.g., production of fertilizers, fossil fuel, and electricity).
- Preservation and enhancement (C sequestration) of carbon stock in agricultural land.
- Reductions of carbon losses from biota and soils.
- Replacement of high CF products with lower CF products for implementing the same services.

Greenhouse gas emissions mitigation has also socioeconomic and political aspects, and the CF of products could be used for product labeling to inform citizens for the best choices they have to reduce the emissions related to food consumption.

4. Social, political, and economic aspects

4.1 Market and demand regulate greenhouse gas emissions

The projected population growth and the nutritional habits that favor higher meat consumption are expected to rise the global demand for food in the coming decades (Henchion et al., 2014; Machovina et al., 2015). This increase in food demand could increase GHGEs, due to intensification of agricultural production and increased cultivation inputs. Reduction of the global food production to mitigate impacts to the global climate system is not considered as a viable solution (Wheeler and von Braun, 2013; Litskas et al., 2019a). A possible reduction of food production impacts food security and increases food market prices, jeopardizing the access to affordable and nutritious food. In addition, during the past decades, companies purchased large agricultural areas in the developing world, exporting the amount of food produced, reducing in many cases the ability of the latter to have access to food of adequate quality and quantity (Collier, 2008; Godfray et al., 2010).

The conversion of natural areas to agricultural land to cover the increasing food demand might have severe impacts on the environment and the climate (Zhang and Cai, 2011). Land use change, especially deforestation, is a significant source for GHGEs and affects other important ecosystem services, such as biodiversity that regulates global ecosystem function (Maes et al., 2016). Healthy ecosystems have the ability to better regulate climate (Maes et al., 2016).

The impact of CC on food production might be more intense in poor countries, with lower ability to adapt due to financial limitations (Hertel et al., 2010). Those located in the tropics or in arid and semiarid areas could experience higher frequency of extreme weather events (e.g., drought or floods; Rojas et al., 2013; Chebbi et al., 2018) and, in some cases, increased impact on crops due to more favorable environment for several notorious pests (Litskas et al., 2019a).

The problem of producing low CF food under global change requires a global policy approach to CC, energy, and food security. The food system needs to transform toward resource efficiency and reduce its GHGEs (Clune et al., 2017). Yields should be safeguarded reducing the dependence on agrochemicals such as pesticides and fertilizers (Litskas et al., 2017; Michos et al., 2018). It is also required to reduce food waste and consumption of high CF foods, such as fish, dairy, and meat (Clune et al., 2017).

4.2 Carbon footprint labeling

Consumers, companies, organizations, and governments are seeking ways to mitigate GHGEs, toward a low carbon economy (Hillier et al., 2011; Norse, 2012). The recent Paris Agreement and the "Roadmap 2050" for low carbon EU economy target emissions reductions up to 80%–95% by sector (ECF, 2019). To achieve the ambitious goals, CF labeling of products is a meaningful indicator and instrument (Vanclay et al., 2011; Hartikainen et al., 2014).

For companies, LCA and CF determination are attractive because the production system could be optimized, in terms of inputs and reduced impacts to climate (Michos et al., 2018). The low CF could be used for advertising products. In addition, input optimization is linked to cost reductions (e.g., less electricity, energy, and material use) (Nelson et al., 2012).

There are examples for the use of a carbon label that describes the CF of a product in countries such as the UK, Australia, and Switzerland. A representative example for the establishment of CF labeling for agricultural products is the initiative of the Cool Farm Alliance (https://coolfarmtool.org/). Its members are academic and research institutions, Non-Govenmental Organizations (NGOs), and representatives of the food industry. Their mission is to provide to the farmers the tools (e.g., simple software) to allow them to report the CF of their products, to communicate it to the merchants and the consumers (Hillier et al., 2011). The use of CF on the labels of products can affect consumers' choices. Vanclay et al. (2011), in a pilot study in Australia, showed that sales of high CF-labeled products decreased.

The Swiss federation of organic farmers Bio Suisse (https://www.bio-suisse.ch/en/home.php) includes the following, in the criteria for awarding its label:

- Products should not be imported into Switzerland by air transport.
- Priority to organic products imports from nearby countries.
- Fresh products (fresh fruit, vegetables, herbs), fruit juices, and frozen products from overseas (except from the Mediterranean) cannot be labeled with the Bio Suisse organic label.

The criteria are designed to support low CF products. Examples of products that have been refused contracts in recent years due to these restrictions are wine and tinned tomatoes from overseas, caviar, and instant iced tea. The organization selected the criteria by following consumers' concerns on environmental issues, including GHGEs produced due to large distance transportation.

5. Beyond carbon footprint, the product environmental footprint

The European Commission (EC) has published a Communication on 'Building the Single Market for Green Products', dated April 9, 2013, with the overall aim of facilitating better information on the environmental performance of products and organizations (European Commission, 2013; http://ec.europa.eu/environment/eussd/smgp/). This communication outlines a framework for the development of methods that would determine the: PEF, or the OEF. Pilots and testing were performed with the participation of a wide range of stakeholders. This testing developed into the Environmental Footprint Rules pilot phase that started at the end of 2013. As an indicative example of a food product, the impact categories that are suggested to be included in the PEF of olive oil are as follows (Schau et al., 2016):

- Carbon footprint
- Ozone depletion
- Ecotoxicity
- Particulate matter emissions
- Acidification
- Eutrophication
- Resource depletion
- Land use

The methodologies and the indicators that are proposed for the determination of the PEF for olive oil are presented in the Product Environmental Footprint Category Rules (PEFCR) for olive oil (Schau et al., 2016). Besides olive oil, other food and drink products have been included in this initiative, such as wine, beer, and feed for animal production.

6. Conclusions

The human influence on the climate system is clear and growing. Climate change affects food production. Energy and fuel consumption as well as other inputs (e.g., fertilizers, materials) for food production, processing, storage, and transportation lead to GHGEs. Intensification of agricultural systems to increase productivity requires increased inputs, but to reduce GHGEs, the food system requires to be transformed toward low inputs and resource efficiency. Life cycle assessment is the method to estimate the CF of food products and allows identification of hotspots to reduce it. Vegetables grown outdoors (e.g., onion, carrot) usually have low CF, whereas the highest is observed for vegetables grown in greenhouses, rice, and meat production. Local varieties that are well adapted to the climate could be a good option for GHGEs mitigation. The use of the CF or the PEF on the labels of products might be a good option to inform consumers about the environmental impact of food products and needs to be promoted.

References

Abeliotis, K., Detsis, & V., Pappia, C., 2013. Life cycle assessment of bean production in the Prespa National Park, Greece. Journal of Cleaner Production 41, 89–96.

Albaladejo, J., Ortiz, R., Garcia-Franco, N., et al., 2013. J Soils Sediments 13, 265. https://doi.org/10.1007/s11368-012-0617-7.

Asseng, S., Martre, P., Maiorano, A., Rötter, R.P., O'Leary, G.J., Fitzgerald, G.J., et al., 2019. Climate change impact and adaptation for wheat protein. Global Change Biology 25 (1), 155–173. https://doi.org/10.1111/gcb.14481.

Carnicer, J., Coll, M., Ninyerola, M., Pons, X., Sanchez, G., Penuelas, J., 2011. Widespread crown condition decline, food web disruption, and amplified tree mortality with increased climate change-type drought. Proceedings of the National Academy of Sciences 108, 1474–1478. https://doi.org/10.1073/pnas.1010070108.

CGIAR, 2019. Direct Agricultural Emissions. Available at: https://ccafs.cgiar.org/bigfacts/#theme=food-emissions&subtheme=direct-agriculture.

Chadwick, D., Sommer, S., Thorman, R., Fangueiro, D., Cardenas, L., Amon, B., Misselbrook, T., 2011. Manure management: implications for greenhouse gas emissions. Animal Feed Science and Technology 166–167, 514–531. https://doi.org/10.1016/j.anifeedsci.2011.04.036. Special Issue: Greenhouse Gases in Animal Agriculture – Finding a Balance between Food and Emissions.

Chebbi, W., Boulet, G., Le Dantec, V., Lili Chabaane, Z., Fanise, P., Mougenot, B., Ayari, H., 2018. Analysis of evapotranspiration components of a rainfed olive orchard during three contrasting years in a semi-arid climate. Agricultural and Forest Meteorology 256–257, 159–178.

Clune, S., Crossin, E., Verghese, K., 2017. Systematic review of greenhouse gas emissions for different fresh food categories. Journal of Cleaner Production 140, 766–783. https://doi.org/10.1016/j.jclepro.2016.04.082.

Collier, P., 2008. The politics of hunger: how illusion and greed fan the food crisis. Foreign Affairs 87, 67–79.

ECF (European Climate Foundation), 2010. Roadmap 2050. Available at: https://www.roadmap2050.eu/.

EEA, 2015. Agriculture and Climate Change. Available at: https://www.eea.europa.eu/signals/signals-2015/articles/agriculture-and-climate-change.

EMEP/EEA, 2009. Air Pollutant Emission Inventory Guidebook. European Environment Agency, Copenhagen. Technical Report No 9.

Eurostat, 2017. GHGE Statistics. Available: https://ec.europa.eu/eurostat/statistics-explained/index.php/Greenhouse_gas_emission_statistics.

European Climate Foundation, 2019. Roadmap 2050. Available at: https://www.roadmap2050.eu/attachments/files/Corepresentation.pdf.

European Commission, 2013. Building the Single Market for Green Products. Available at: http://ec.europa.eu/environment/eussd/smgp/.

Flessa, H., Ruser, R., Dörsch, P., Kamp, T., Jimenez, M.A., Munch, J.C., Beese, F., 2002. Integrated evaluation of greenhouse gas emissions (CO_2, CH_4, N_2O) from two farming systems in southern Germany. Agriculture, Ecosystems and Environment 91 (1–3), 175–189.

Godfray, H.C.J., Beddington, J.R., Crute, I.R., Haddad, L., Lawrence, D., Muir, J.F., Pretty, J., Robinson, S., Thomas, S.M., Toulmin, C., 2010. Food security: the challenge of feeding 9 billion people. Science 327, 812–818. https://doi.org/10.1126/science.1185383.

Hartikainen, H., Roininen, T., Katajajuuri, J.-M., Pulkkinen, H., 2014. Finnish consumer perceptions of carbon footprints and carbon labelling of food products. Journal of Cleaner Production 73, 285–293. https://doi.org/10.1016/j.jclepro.2013.09.018.

Henchion, M., McCarthy, M., Resconi, V.C., Troy, D., 2014. Meat consumption: trends and quality matters. Meat Science 98, 561–568. https://doi.org/10.1016/j.meatsci.2014.06.007.

Herrero, M., Havlik, P., Valin, H., Notenbaert, A., Rufino, M.C., Thornton, P.K., Blummel, M., Weiss, F., Grace, D., Obersteiner, M., 2013. Biomass use, production, feed efficiencies, and greenhouse gas emissions from global livestock systems. Proceedings of the National Academy of Sciences 110, 20888–20893. https://doi.org/10.1073/pnas.1308149110.

Herrero, M., Henderson, B., Havlík, P., Thornton, P.K., Conant, R.T., Smith, P., Wirsenius, S., Hristov, A.N., Gerber, P., Gill, M., Butterbach-Bahl, K., Valin, H., Garnett, T., Stehfest, E., 2016. Greenhouse gas mitigation potentials in the livestock sector. Nature Climate Change 6, 452–461. https://doi.org/10.1038/nclimate2925.

Hertel, T.W., Burke, M.B., Lobell, D.B., 2010. The poverty implications of climate-induced crop yield changes by 2030. Global Environmental Change 20, 577–585. https://doi.org/10.1016/j.gloenvcha.2010.07.001.

Hillier, J., Walter, C., Malin, D., Garcia-Suarez, T., Mila-i-Canals, L., Smith, P., 2011. A farm-focused calculator for emissions from crop and livestock production. Environmental Modelling and Software 26, 1070–1078. https://doi.org/10.1016/j.envsoft.2011.03.014.

Hillier, J., Brentrup, F., Wattenbach, M., Walter, C., Garcia-Suarez, T., Mila-i-Canals, L., Smith, P., 2012. Which cropland greenhouse gas mitigation options give the greatest benefits in different world regions? Climate and soil-specific predictions from integrated empirical models. Global Change Biology 18 (6), 1880–1894.

Hoegh-Guldberg, O., Jacob, D., Taylor, M., Bindi, M., Brown, S., Camilloni, I., Diedhiou, A., Djalante, R., Ebi, K.L., Engelbrecht, F., Guiot, J., Hijioka, Y., Mehrotra, S., Payne, A., Seneviratne, S.I., Thomas, A., Warren, R., Zhou, G., 2018. Impacts of 1.5 oC global warming on natural and human systems. In: Masson-Delmotte, V., Zhai, P., Pörtner, H.-O., Roberts, D., Skea, J., Shukla, P.R., Pirani, A., Moufouma-Okia, W., Péan, C., Pidcock, R., Connors, S., Matthews, J.B.R., Chen, Y., Zhou, X., Gomis, M.I., Lonnoy, E., Maycock, T., Tignor, M., Waterfield, T. (Eds.), Global Warming of 1.5°C. An IPCC Special Report on the Impacts of Global Warming of 1.5 °C above Pre-industrial Levels and Related Global Greenhouse Gas Emission Pathways, in the Context of Strengthening the Global Response to the Threat of Climate Change, Sustainable Development, and Efforts to Eradicate Poverty (in press).

Hoeppner, J.W., Entz, M.H., McConkey, B.G., Zentner, R.P., Nagy, C.N., 2006. Energy use and efficiency in two Canadian organic and conventional crop production systems. Renewable Agriculture and Food Systems 21 (1), 60–67.

Hou, Y., Velthof, G.L., Oenema, O., 2015. Mitigation of ammonia, nitrous oxide and methane emissions from manure management chains: a meta-analysis and integrated assessment. Global Change Biology 21, 1293–1312. https://doi.org/10.1111/gcb.12767.

Iglesias, A., Garrote, L., 2015. Adaptation strategies for agricultural water management under climate change in Europe. Agricultural Water Management 155, 113–124. https://doi.org/10.1016/j.agwat.2015.03.014.

IPCC, 1997. Greenhouse Gas Inventory Reference Manual, Three Volumes, Intergovernmental Panel on Climate Change. IPCC Technical Support Unit, London.

IPCC, 2006. Software for National Gas Inventories, Intergovernmental Panel on Climate Change. http://ipcc2006.air.sk/.

IPCC, 2014. In: Edenhofer, O., Pichs-Madruga, R., Sokona, Y., Farahani, E., Kadner, S., Seyboth, K., Adler, A., Baum, I., Brunner, S., Eickemeier, P., Kriemann, B., Savolainen, J., Schlömer, S., von Stechow, C., Zwickel, T., Minx, J.C. (Eds.), Climate Change 2014: Mitigation of Climate Change. Contribution of Working Group III to the Fifth Assessment Report of the Intergovernmental Panel on Climate Change. Cambridge University Press, Cambridge, United Kingdom and New York, NY, USA.

ISO, 2006a. Environmental Management-Life Cycle Assessment-Principles and Framework ISO-Norm 14040:2006. International Organization for Standardization. European Committee for Standardization, Brussels, Belgium.

ISO, 2006b. Environmental Management-Life Cycle Assessment-Requirements and Guidelines ISO-Norm 14044:2006. International Organization for Standardization. European Committee for Standardization, Brussels, Belgium.

ISO, 2013. Carbon Footprint of Products—Requirements and Guidelines for Quantification and Communication. ISONorm 14067:2013. International Organization for Standardization. European Committee for Standardization, Brussels, Belgium. http://www.iso.org/iso/iso_catalogue/catalogue_tc/catalogue_detail.htm?csnumber=59521.

Kavargiris, S.E., Mamolos, A.P., Tsatsarelis, C.A., Nikolaidou, A.E., Kalburtji, K.L., 2009. Energy resources' utilization in organic and conventional vineyards: energy flow, greenhouse gas emissions and biofuel production. Biomass and Bioenergy 33, 1239–1250. https://doi.org/10.1016/j.biombioe.2009.05.006.

Kaltsas, A.M., Mamolos, A.P., Tsatsarelis, C.A., Nanos, G.D., Kalburtji, K.L., 2007. Energy budget in organic and conventional olive groves. Agriculture, Ecosystems and Environment 122 (2), 243–251.

Kehagias, M.C., Michos, M.C., Menexes, G.C., Mamolos, A.P., Tsatsarelis, C.A., Anagnostopoulos, C.D., Kalburtji, K.L., 2015. Energy equilibrium and carbon dioxide, methane, and nitrous oxide-emissions in organic, integrated and conventional apple orchards related to Natura 2000 site. Journal of Cleaner Production 91, 89–95. https://doi.org/10.1016/j.jclepro.2014.12.007.

Litskas, D.V., Migeon, A., Navajas, M., Tixier, M.-S., Stavrinides, C.M., 2019a. Impacts of climate change on tomato, a notorious pest and its natural enemy: small scale agriculture at higher risk. Environmental Research Letters 14 (8). https://doi.org/10.1088/1748-9326/ab3313.

Litskas, D.V., Chrysargyris, A., Stavrinides, C.M., Tzortzakis, N., 2019b. Water-energy-food nexus: a case study on medicinal and aromatic plants. Journal of Cleaner Production 233 (1). https://doi.org/10.1016/j.jclepro.2019.06.065.

Litskas, V.D., Irakleous, T., Tzortzakis, N., Stavrinides, M.C., 2017. Determining the carbon footprint of indigenous and introduced grape varieties through life cycle assessment using the island of Cyprus as a case study. Journal of Cleaner Production 156, 418–425. https://doi.org/10.1016/j.jclepro.2017.04.057.

Litskas, V.D., Karaolis, C.S., Menexes, G.C., Mamolos, A.P., Koutsos, T.M., Kalburtji, K.L., 2013. Variation of energy flow and greenhouse gas emissions in vineyards located in Natura 2000 sites. Ecological Indicators 27, 1–7. https://doi.org/10.1016/j.ecolind.2012.11.016.

Litskas, V.D., Mamolos, A.P., Kalburtji, K.L., Tsatsarelis, C.A., Kiose-Kampasakali, E., 2011. Energy flow and greenhouse gas emissions in organic and conventional sweet cherry orchards located in or close to natura 2000 sites. Biomass and Bioenergy 35, 1302–1310. https://doi.org/10.1016/j.biombioe.2010.12.023.

Liu, Y., Wang, E., Yang, X., Wang, J., 2009. Contributions of climatic and crop varietal changes to crop production in the North China Plain, since 1980s: crop yield impacted by crop variety and climate. Global Change Biology 16, 2287–2299. https://doi.org/10.1111/j.1365-2486.2009.02077.x.

Machovina, B., Feeley, K.J., Ripple, W.J., 2015. Biodiversity conservation: the key is reducing meat consumption. The Science of the Total Environment 536, 419–431. https://doi.org/10.1016/j.scitotenv.2015.07.022.

Maes, J., Zulian, G., Thijssen, M., Castell, C., Baró, F., Ferreira, A.M., Melo, J., Garrett, C.P., David, N., Alzetta, C., Geneletti, D., Cortinovis, C., Zwierzchowska, I., Louro Alves, F., Souto Cruz, C., Blasi, C., Alós Ortí, M.M., Attorre, F., Azzella, M.M., Capotorti, G., Copiz, R., Fusaro, L., Manes, F., Marando, F., Marchetti, M., Mollo, B., Salvatori, E., Zavattero, L., Zingari, P.C., Giarratano, M.C., Bianchi, E., Duprè, E., Barton, D., Stange, E., Perez-Soba, M., Eupen, M. van, Verweij, P., Vries, A. de, Kruse, H., Polce, C., Cugny-Seguin, M., Erhard, M., Nicolau, R., Fonseca, A., Fritz, M., Teller, A., European Commission, Directorate-General for the Environment, 2016. In: Mapping and Assessment of Ecosystems and Their Services: Urban Ecosystems, 4th Report – Final, May 2016. Publications Office, Luxembourg.

Martin, C., Morgavi, D.P., Doreau, M., 2010. Methane mitigation in ruminants: from microbe to the farm scale. Animal 4 (3), 351–365.

Michos, M.C., Mamolos, A.P., Menexes, G.C., Tsatsarelis, C.A., Tsirakoglou, V.M., Kalburtji, K.L., 2012. Energy inputs, outputs and greenhouse gas emissions in organic, integrated and conventional peach orchards. Ecological Indicators 13, 22–28.

Michos, M.C., Menexes, G.C., Kalburtji, K.L., Tsatsarelis, C.A., Anagnostopoulos, C.D., Mamolos, A.P., 2017. Could energy flow in agro-ecosystems be used as a "tool" for crop and farming system replacement? Ecological Indicators 73, 247–253.

Michos, M.C., Menexes, G.C., Mamolos, A.P., Tsatsarelis, C.A., Anagnostopoulos, C.D., Tsaboula, A.D., Kalburtji, K.L., 2018. Energy flow, carbon and water footprints in vineyards and orchards to determine environmentally favourable sites in accordance with Natura 2000 perspective. Journal of Cleaner Production 187, 400–408. https://doi.org/10.1016/j.jclepro.2018.03.251.

Mohammad, W., Shah, S.M., Shehzadi, S., Shah, S.A., 2012. Effect of tillage, rotation and crop residues on wheat crop productivity, fertilizer nitrogen and water use efficiency and soil organic carbon status in dry area (rainfed) of north-west Pakistan. Journal of Soil Science and Plant Nutrition. https://doi.org/10.4067/S0718-95162012005000027.

Nelson, J., Johnston, J., Mileva, A., Fripp, M., Hoffman, I., Petros-Good, A., Blanco, C., Kammen, D.M., 2012. High-resolution modeling of the western North American power system demonstrates low-cost and low-carbon futures. Energy Policy 43, 436–447. https://doi.org/10.1016/j.enpol.2012.01.031.

Norse, D., 2012. Low carbon agriculture: objectives and policy pathways. Environmental Development 1, 25–39. https://doi.org/10.1016/j.envdev.2011.12.004.

Olesen, J.E., Schelde, K., Weiske, A., Weisbjerg, M.R., Asman, W.A., Djurhuus, J., 2006. Modelling greenhouse gas emissions from European conventional and organic dairy farms. Agriculture, Ecosystems and Environment 112 (2–3), 207–220.

Pergola, M., Persiani, A., Pastore, V., Palese, A.M., Arous, A., Celano, G., 2017. A comprehensive Life Cycle Assessment (LCA) of three apricot orchard systems located in Metapontino area (Southern Italy). Journal of Cleaner Production 142 (4), 4059–4071.

Ponti, L., Gutierrez, A.P., Ruti, P.M., Dell'Aquila, A., 2014. Fine-scale ecological and economic assessment of climate change on olive in the Mediterranean Basin reveals winners and losers. Proceedings of the National Academy of Sciences 111, 5598–5603. https://doi.org/10.1073/pnas.1314437111.

Popp, A., Lotze-Campen, H., Bodirsky, B., 2010. Food consumption, diet shifts and associated non-CO_2 greenhouse gases from agricultural production. Global Environmental Change 20 (3), 451–462.

Popp, A., Humpenöder, F., Weindl, I., Bodirsky, B.L., Bonsch, M., Lotze-Campen, H., et al, 2014. Land-use protection for climate change mitigation. Nature Climate Change 4 (12), 1095.

Rojas, R., Feyen, L., Watkiss, P., 2013. Climate change and river floods in the European Union: socio-economic consequences and the costs and benefits of adaptation. Global Environmental Change 23, 1737–1751. https://doi.org/10.1016/j.gloenvcha.2013.08.006.

Rötter, R.P., Palosuo, T., Kersebaum, K.C., Angulo, C., Bindi, M., Ewert, F., Ferrise, R., Hlavinka, P., Moriondo, M., Nendel, C., Olesen, J.E., Patil, R.H., Ruget, F., Takáč, J., Trnka, M., 2012. Simulation of spring barley yield in different climatic zones of Northern and Central Europe: a comparison of nine crop models. Field Crops Research 133, 23–36. https://doi.org/10.1016/j.fcr.2012.03.016.

Savary, S., Willocquet, L., Pethybridge, S.J., Esker, P., McRoberts, N., Nelson, A., 2019. The global burden of pathogens and pests on major food crops. Nature Ecology and Evolution 3 (3), 430–439. https://doi.org/10.1038/s41559-018-0793-y.

Sadras, V.O., Lawson, C., 2013. Nitrogen and water-use efficiency of Australian wheat varieties released between 1958 and 2007. European Journal of Agronomy 46, 34–41. https://doi.org/10.1016/j.eja.2012.11.008.

Schau, E.M., Polo Palomino, J.A., Michalopoulos, G., Russo, C., 2016. Product Environmental Footprint Category Rules for Olive Oil – 3rd Draft. Available at: https://ec.europa.eu/environment/eussd/smgp/pdf/pilots/draft_pefcr_olive_oil_pilot_for_3rd_consultation.pdf.

Smith, P., Haberl, H., Popp, A., Erb, K.H., Lauk, C., Harper, R., Tubiello, F.N., de Siqueira Pinto, A., Jafari, M., Sohi, S., Masera, O., 2013. How much land-based greenhouse gas mitigation can be achieved without compromising food security and environmental goals? Global Change Biology 19 (8), 2285–2302.

Stehfest, E., Bouwman, L., Van Vuuren, D.P., Den Elzen, M.G., Eickhout, B., Kabat, P., 2009. Climate benefits of changing diet. Climatic Change 95 (1–2), 83–102.

Teixeira, E.I., Fischer, G., van Velthuizen, H., Walter, C., Ewert, F., 2013. Global hot-spots of heat stress on agricultural crops due to climate change. Agricultural and Forest Meteorology 170, 206–215. https://doi.org/10.1016/j.agrformet.2011.09.002.

Vanclay, J.K., Shortiss, J., Aulsebrook, S., Gillespie, A.M., Howell, B.C., Johanni, R., Maher, M.J., Mitchell, K.M., Stewart, M.D., Yates, J., 2011. Customer response to carbon labelling of groceries. Journal of Consumer Policy 34, 153–160. https://doi.org/10.1007/s10603-010-9140-7.

Wheeler, T., von Braun, J., 2013. Climate change impacts on global food security. Science 341, 508–513. https://doi.org/10.1126/science.1239402.

World Bank, 2019. Agriculture, Forestry, And Fishing, Value Added (Current US$). https://data.worldbank.org/indicator/NV.AGR.TOTL.CD?end=2017&start=1980.

Zhang, X., Cai, X., 2011. Climate change impacts on global agricultural land availability. Environmental Research Letters 6, 014014. https://doi.org/10.1088/1748-9326/6/1/014014.

Zhou, J., Wang, C., Zhang, H., Dong, F., Zheng, X., Gale, W., Li, S., 2011. Effect of water saving management practices and nitrogen fertilizer rate on crop yield and water use efficiency in a winter wheat–summer maize cropping system. Field Crops Research 122, 157–163. https://doi.org/10.1016/j.fcr.2011.03.009.

PART

Food production and food waste valorization

II

CHAPTER

Valorization of food processing by-products via biofuel production

4

T.A. Trabold, B.S., PhD [1], D. Rodríguez Alberto, B.S., M.S. [2]

[1]*Department Head, Golisano Institute for Sustainability, Rochester Institute of Technology, Rochester, NY, United States;* [2]*PhD Candidate, Golisano Institute for Sustainability, Rochester Institute of Technology, Rochester, NY, United States*

1. Introduction

The problem of food waste has become the focus of intense global scrutiny, because of the significant nutrient, energy, and water resources required to feed the world's growing population, and the environmental impacts associated with moving food materials from primary production to consumption or "farm-to-fork" (Fig. 4.1). Many governmental and nonprofit organizations have become active in addressing the seemingly absurd state of affairs in which 30% or more of food produced globally is never consumed by humans, whereas 27% of the human population suffers from some level of food insecurity (ReFED, 2016; Smith et al., 2017). Groups active

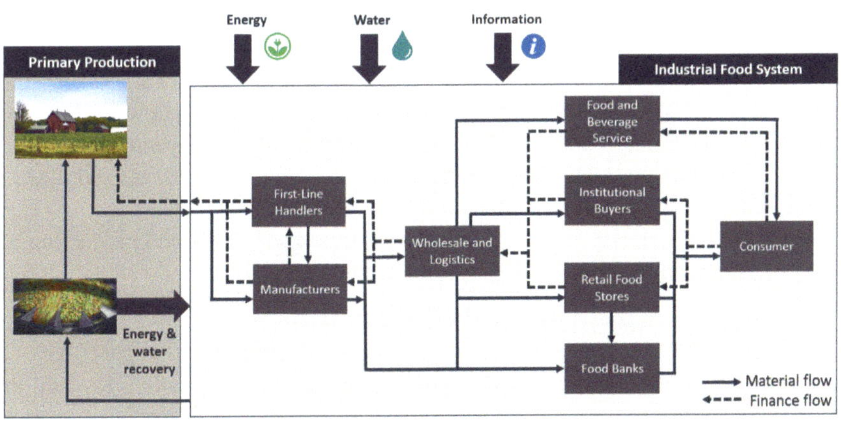

FIGURE 4.1

Material and finance flows in the global food system, also including critical inputs of energy, water, and information. Many opportunities exist for energy and water recovery that can benefit primary production and the industrial food system.

Adapted from National Research Council, 2015. A Framework for Assessing Effects of the Food System. National Academies Press.

Sustainability of the Food System. https://doi.org/10.1016/B978-0-12-818293-2.00004-5

in promoting alternatives to conventional food waste treatment methods (landfills, wastewater treatment, incineration) include the European Institute of Innovation and Technology—Food (EIT Food) in the European Union, The Waste and Resources Action Programme (WRAP) in the United Kingdom, and Rethink Food Waste through Economics and Data (ReFED) in the United States, as well as the Food and Agriculture Organization (FAO) of the United Nations.

Different regions of the global food system have distinctly different profiles of food waste generation (Trabold et al., 2018). Generally speaking, developing economies experience more loss in the agriculture and primary production stages of the food supply chain, largely due to a lack of storage and transportation infrastructure. Conversely, more economically affluent regions generate the majority of their food waste near the consumption part of the supply chain. Food processing companies (or "Manufacturers" as indicated in Fig. 4.1) operate at the intersection between primary production and the industrial food system and, unlike most other stages, contribute a remarkably consistent fraction of total food waste generated (on a kcal basis) across all global regions (Lipinski et al., 2013):

- Industrialized Asia: 2%
- South and Southeast Asia: 4%
- North Africa, West and Central Asia: 4%
- Europe: 5%
- Latin America: 6%
- Sub-Saharan Africa: 7%
- North America and Oceania: 9%

These relatively modest waste generation fractions would seem to imply that food processing operations are efficient in how they utilize primary products from agricultural operations and prepare these materials for entry into the downstream supply chain, starting with wholesale and logistics. In this sense, other stakeholders across the global food system may benefit by better understanding how food waste is managed in processing plants that can maintain economic viability and competitiveness only if all system inputs and outputs are controlled, optimized, and valorized to the greatest possible extent.

Aside from the strong economic incentives driving efficient food processing operations, this stage benefits from two other important factors:

- Processing plants generally manufacture a small number of similar products, and thus, any waste produced is relatively homogeneous and contaminant-free.
- Waste is generated in a few locations at relatively large per-site rates, and this offers economies of scale for a variety of waste valorization strategies, such as those shown in Fig. 4.2.

Although many prior studies have documented the use of food processing wastes in animal feed, compost, and other value-added products (e.g., Lin et al., 2013), the focus of this chapter is on using these materials for biofuels production based on the technologies outlined in Section 2. Generally speaking, biofuels cannot provide as

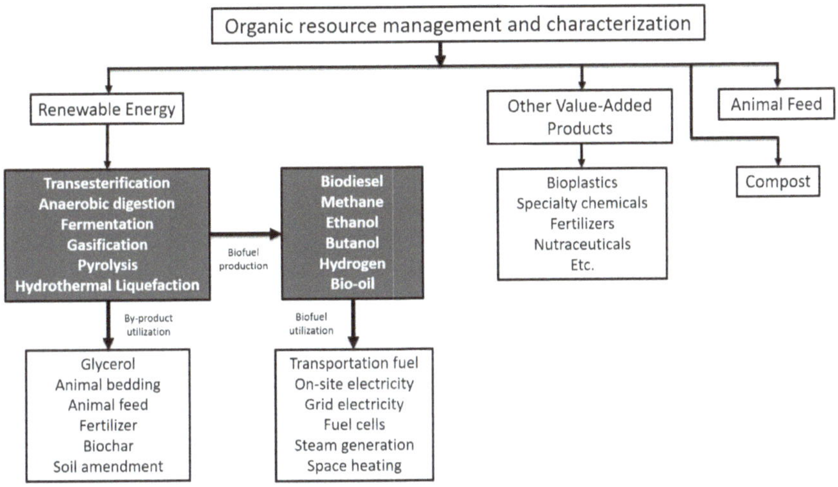

FIGURE 4.2

Flowchart of possible food waste valorization pathways, with biofuel production as a practical option for food processing wastes that are heterogeneous or contain contaminants, making them unsuitable for higher-value products such as nutraceuticals or animal feed.

great an economic rate of return (on a per kg waste basis) as many other value-added products but may have utility in cases where the organic matter is fairly dilute or generated at a relatively low volume, so combining it with other available waste resources makes economic sense. Upon mixing wastes with different properties, many of the higher-value valorization options are lost, but such combinations may serve quite well as biofuel feedstocks. In fact, biofuels are already produced from food waste on a massive scale around the world, mostly through the process of anaerobic digestion (AD) in landfills and wastewater treatment plants. These conventional waste management systems, however, are generally not efficient converters of the embodied energy present in food waste and, at least in the case of landfills, do not sufficiently capture the generated methane to avoid its deleterious impact on global warming. A much more economically and environmentally sustainable approach is to develop systems that intentionally collect and convert food waste to energy, thereby enabling the food processing waste generator to benefit from the material they invested in from the start.

2. Methods for production of biofuels from food processing by-products

As stated above, the production of biofuels from food processing wastes is not necessarily the most economically or environmentally sustainable option, and other valorization pathways may offer better return on investment to the waste generator. However, in many cases, biofuel production makes sense for waste streams with inconsistent physical and/or chemical properties, containing small amounts of impurities that render them unsuitable for upcycling to secondary food products for humans or animals, or are available in a diluted state making combination with other organic waste streams (e.g., dairy manure) the most practical strategy. Also, transportation logistics play a central role in determining when food waste-to-biofuel conversion is viable. For example, a food processing plant may elect to divert waste to an anaerobic digester located in close proximity, even with payment of a tipping fee, instead of taking on the development and management costs of converting the same material to a secondary food product that could generate another revenue stream. This outcome is often the case with smaller, regional facilities that do not have significant in-house research and development expertise, or lack the desire to expand beyond their core business areas.

In the following discussion, we present three main conversion technologies that have been previously applied to convert food processing wastes into biofuels:

- *AD* to produce hydrogen- or methane-rich biogas
- *Fermentation* to produce liquid alcohol fuels, such as ethanol and butanol, as well as hydrogen
- *Thermochemical conversion (TC)*, including gasification, pyrolysis, and hydrothermal liquefaction (HTL), to produce hydrogen-rich syngas and bio-oil

Transesterification is another widely used process to convert waste vegetable oil to fatty acid methyl esters, commonly known as biodiesel. However, because this technology typically uses a single feedstock material, it is not discussed in detail here and the reader is referred to available reference resources (e.g., Van Gerpen, 2005; Knothe et al., 2015). Beyond describing the three core technologies above and reviewing the recent literature relevant to this chapter, we do not consider whether or not the specific technology or process methodology has achieved commercial scale, nor do we consider the various ways in which the biofuels thus derived can be productively utilized for electricity generation, transportation fuel, combined heat and power (CHP), etc. The potential uses of by-products of biofuel production (e.g., glycerol, fertilizer, biochar) and different possible scales of operation are discussed in connection with the waste biorefinery concept in Section 3. It should be noted at the outset that there are a number of useful reference resources that have previously addressed various aspects of food waste valorization to biofuels, including Arvanitoyannis (2010), Chandrasekaran (2013), Kosseva and Webb (2013), de Jong and van Ommen (2014), and Trabold and Babbitt (2018).

2.1 Anaerobic digestion

AD is a multistep biochemical process whereby microorganisms break down degradable organic matter in the absence of oxygen. It is a common method for converting food waste into renewable biogas, composed mostly of methane (CH_4) and carbon dioxide (CO_2), combined with other minor constituents such as water vapor and hydrogen sulfide (H_2S) present in much smaller quantities.

AD systems are deployed worldwide, from the scale of small family-sized systems numbering in the millions in China and India, up to very large industrial systems concentrated in Europe.

Because of the importance of AD as an industrial waste management technology, many papers have provided overviews of different system configurations, operating parameters, and biogas production potential of various food waste materials, including several reviews published in the past year: Labatut and Pronto (2018), Li et al. (2018), Ren et al. (2018), Xu et al. (2018). From these reviews, it is apparent that most of the published studies cover bench-scale experiments with volumes of 10 L or less and often do not comprehend key control parameters that are critical to operation at full scale, such as the organic loading rate (ORL, the quantity of digestible feedstock entering the AD reactor per unit time) and the carbon-to-nitrogen ratio (C/N) of combined feedstocks, the latter typically recommended to be controlled in the range of 20–30 (Labatut and Pronto, 2018). For a particular food waste material, a combination of various food waste materials, or food mixed with manure, the rate of biogas production would be expected to increase with increasing OLR up a certain optimal level dictated by the specific chemical and micronutrient properties of the substrates undergoing digestion. It is also apparent that the potential of AD as a sustainable food waste management strategy relies heavily on the specific types of food waste available. For example, based on the results of the study of Ebner et al. (2016) and further analysis by Labatut and Pronto (2018), the average volumes of biomethane produced (at standard temperature and pressure) by salad mixes, ice cream, spent coffee grounds and filters, and stale bakery products were 14, 55, 106, and 414 L of CH_4 per kg of waste, respectively. Generally, materials with high lipid and/or degradable carbohydrate contents produced the greatest amount of methane. Ebner et al. (2016) also showed that some wastes can be combined in a synergistic way to produce more biogas than can be generated by the same substrates processed individually, with up to a 20% enhancement when weighted on a volatile solids (VS) basis.

Despite the availability of extensive literature on the biomethane potential of a wide variety of food waste materials, in practice it is challenging to operate "pure food waste" digesters unless processing conditions are optimized for specific properties, OLR, hydraulic retention time (HRT), etc. (Fiore et al., 2016). In many regions, anaerobic codigestion (AcoD) is the preferred practice, whereby a stable primary substrate (such as dairy manure) is combined with mixed secondary substrates (including food wastes) that may vary significantly over time in regard to their chemical/physical properties and influent volumes. To achieve high ORL and maintain process stability, many AcoD plants operate with primary-to-secondary substrate ratios of 70:30 and usually not in excess of 50:50. Codigestion has the benefit of greatly enhancing biogas production for relatively low-energy substrates such as wastewater sludge or animal manure (Liu et al., 2016), and it has been reported that codigestion yields significant reductions in net greenhouse gas (GHG) emissions relative to conventional manure and food waste management practices (Ebner et al., 2015; Usack et al., 2018). In regard to AcoD plant operation, it is instructive to understand the actual profile of the types of food waste materials employed at commercial scale. As an example of data acquired from a large facility operating on a dairy farm in upstate New York with a 1.4-MW engine generator set, Fig. 4.3 shows average monthly waste volumes and delivery frequencies over a 6-month period for 17 industrial food system sources. The data reveal very large variations in the volumes of waste provided from individual entities, from over 1.6 million L/month to less than 4000 L/month, and delivery frequencies ranging from 26 to 1 per month. It is also important to note that all waste sources identified in Fig. 4.3 are in liquid or semiliquid phase and have few alternative valorization options available, such as composting that can be used for many solid phase pre- and postconsumer waste streams. The advantage of using liquid wastes is that there is minimal pretreatment (such as grinding) required to make the material suitable for pumping through the AD system.

2.2 Fermentation

Fermentation relates to a variety of biochemical processes facilitated by microbes (e.g., *Saccharomyces cerevisiae*), and in the context of this chapter, it is used to refer to processes generating fuels that can potentially be used for sustainable transportation or other energy conversion systems based on ethanol, butanol, and hydrogen. It is well known that ethanol is produced at very large scale using primarily corn in the United States, sugar cane in Brazil, and other dedicated energy crops in Europe and

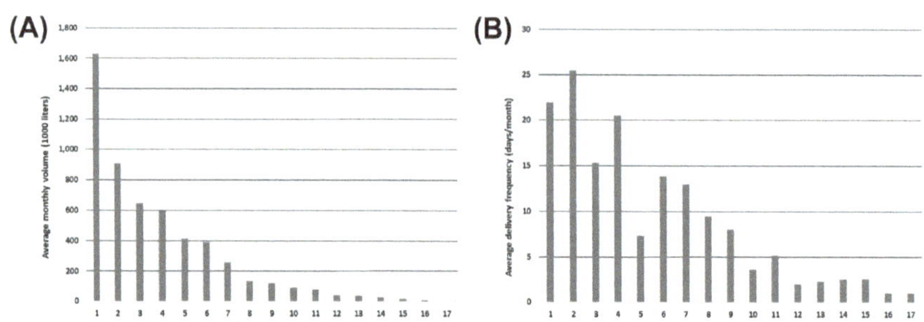

(C)

Source	Description	Source	Description
1	Cheese process wash and over production	10	FOG from grease traps
2	Organic milk waste	11	Rice drink & rice yogurt production effluents
3	Dairy permeate, DAF and production waste	12	Syrup and sauce overproduction
4	FOG	13	Yellow grease from restaurants
5	Unsold skim and butter milk	14	FOG from grease traps
6	Dairy whey	15	FOG from grease traps
7	Whipped topping waste	16	FOG from grease traps
8	FOG from grease traps	17	FOG from grease traps
9	Yellow grease from restaurants		

*FOG = fats, oils and grease; DAF = dissolved air floatation

FIGURE 4.3

Representative average monthly inputs of food processing wastes to an anaerobic codigestion plant in New York State.

Asia. Although agricultural residues and other cellulosic resources are gaining a greater share of the feedstock mix, food processing wastes account for a very small fraction of the materials used for ethanol production, and facilities using food waste are generally much smaller than those using conventional commodity crops that often produce more than 100 million gallons/year (379 million liters/year). Using data from September 2018, *Ethanol Magazine* listed only six plants in the United States using crop residues or food waste streams for ethanol production[1]: DuPont Denisco (Nevada, IA, 114 MMLy); Merrick/Coors (Golden, CO, 11 MMLy); Parallel Products (Rancho Cucamonga, CA, 6 MMLy); Parallel Products (Louisville, KY, 23 MMLy); Poet (Emmetsburg, IA, 76 MMLy); and Summit Natural Energy (Cornelius, OR, 4 MMLy). Butanol is a four-carbon alcohol with several features that make it a potentially better transportation fuel and closer drop-in replacement for gasoline in internal combustion engines (ICEs). These features include higher energy content, lower vapor pressure, lower water miscibility, and lower corrosivity than ethanol (Harvey and Meylemans, 2011). Hydrogen fuel can also be produced by a number of different processes, with dark fermentation being the most commercially viable due to no requirement for external energy input or lighting and generally lower operating cost (Yasin et al., 2013; Łukajtis et al., 2018).

Recent papers by Hegde et al. (2018) and Hegde and Trabold (2018) have provided comprehensive reviews of published studies describing the use of food processing waste streams for production of ethanol, butanol, and hydrogen:

- Ethanol → bakery waste, waste potato mash, potato peels, sweet potato waste, apple pomace, grape pomace, tomato serum from sauce production, cheese whey permeate, pineapple, and banana peels
- Butanol → bakery waste, waste potato, wastewater from palm oil production, cheese whey, apple pomace, and acid whey
- Hydrogen → tofu processing wastewater, cheese whey, apple pomace, pineapple, banana, and mixed fruit peel waste

Between the two liquid fuels, ethanol yields are typically much higher than butanol, but in both cases, the measured alcohol production rates are generally below levels needed to achieve commercial viability. For ethanol, Hegde et al. (2018) showed that only high solids content materials such as waste bread, restaurant and cafeteria waste, and apple pomace can achieve the ~10% product concentration common in corn ethanol production. For butanol, a concentration of at least 1.5% is considered necessary to have any possibility for scale-up, based on the 2% maximum yields reported for batch fermentation with *Clostridium* strains. Similarly, a small subset of the investigated substrates was determined to be viable for conversion to butanol (only apple pomace and potato starch). Higher yields are possible, but only with a product recovery subsystem that would add processing

[1] http://www.ethanolproducer.com/plants/listplants/US/Operational/All; MMLy = million liters per year.

costs. Although Hegde et al. (2018) suggested that nonfuel solvent alcohols may be higher value products for fermentation of food processing wastes, if energy is the desired outcome, AD (Section 2.1) may be a better option, especially for liquid-phase waste streams.

There may be more potential for the valorization of food processing wastes through the dark fermentation pathway for hydrogen production, because of the current interest in expanding options that move the industry away from conventional steam methane reforming. Using renewable hydrogen in fuel cell vehicles running at much high efficiencies than ICEs is a compelling objective, but much more work is needed to understand the compatibility and integration of food waste–derived hydrogen with proton exchange membrane fuel cell systems used for vehicle transport (del Campo et al., 2012; Rahman et al., 2015).

2.3 Thermochemical conversion

For solid-phase food processing waste streams with relatively low moisture content, TC is a potential valorization option. In the context of this chapter, TC refers to processes based on high-temperature treatment in the complete absence of oxygen (pyrolysis) or with less oxygen than the stoichiometric level needed for full combustion (gasification). Although TC technologies have been widely studied as a pathway for conversion of municipal solid waste (e.g., Sørum et al., 2001; Arena, 2012), there is relatively little research reported on quantifying energy production potential of food waste materials. A recent review article by Guran (2018) summarized studies of animal- and plant-based food materials that have been converted via TC methods, and showed the potential for the valorization of waste streams that may not be suitable for the biochemical technologies described above (Sections 2.1 and 2.2). Such challenging feedstocks include animal bones and carcasses, tree nut and coconut shells, and corncobs. In cases where the desired process output is solely biofuel for subsequent conversion to electrical and/or thermal energy (or vehicle fuel after further purification), gasification is the preferred pathway, because the primary process output is syngas comprised of about 85% hydrogen and carbon dioxide. Ahmed and Gupta (2010) evaluated pyrolysis and gasification using dog food as a model food waste material, and found that the latter provided significantly higher hydrogen yield, but required longer reaction time. Higher hydrogen yield can also be achieved by applying a secondary water-gas shift reaction stage ($CO + H_2O \rightarrow CO_2 + H_2$). However, the performance of a food waste-to-energy gasification process is significantly affected by the specific chemical and physical properties of the waste feedstock, including moisture and volatile matter content, bulk density and size, elemental composition, lower heating value (LHV), inorganic components (some of which can serve as catalysts), and other contaminants (N, S, Cl, etc.) that could result in undesired gas phase constituents (Zevenhoven-Onderwater et al., 2001). Because of the challenges associated with managing TC processes for food waste feedstocks that may vary greatly over time, “there are almost no gasification/pyrolysis processes that have been solely developed for food waste” (Pham et al., 2015).

Gasification is clearly the preferred technology in cases where syngas production is the most economically favorable process output, with only a small amount of ash generated as the other reaction product. On the other hand, pyrolysis has the advantage of providing three main value-added co-products (syngas, bio-oil and solid "char"), the relative quantities of which can be adjusted by judicious control of processing parameters. In *fast pyrolysis*, organic matter is rapidly heated to temperatures around 500°C and the resulting vapors are condensed to produce bio-oil, typically at a 60%–75% yield on a dry basis. A significant advantage of this approach is that the resulting high energy content liquid can be readily stored and transported, thereby decoupling the ultimate application from the production process. The bio-oil can be refined for use as transportation fuel, feedstock for electric power or steam generation, as well as for more advanced applications as raw materials in the production of fertilizers, building materials, phenolic compounds, etc. (Venderbosch and Prins, 2010). In *slow pyrolysis* the organic waste material is heated at a slower ramp rate, and the maximum temperature dictates the relative yields of syngas and solid char, comprised mostly of a very stable form carbon; in this process option bio-oil is typically not the desired product. Grycová et al. (2016) studied the pyrolysis generated syngas composition of several low moisture content waste materials (cereals and peanut crisps), and determined that hydrogen concentrations in excess of 60% were achieved for reaction temperatures in the range of 750–800°C. Lower temperatures may be favorable to increase the yield of char, which when intended for use in soil amendment, environmental management or similar applications is usually referred to as "biochar" (Xu et al., 2011; Lehmann and Joseph, 2015).

TC may be the preferred option when solid waste characteristics are known to vary greatly over time or if there is the possibility of small amounts of contamination that could adversely impact living organisms responsible for the fundamental biochemical processes in AD and fermentation systems. For example, pyrolysis may be suitable for "real-world" waste streams where it is impractical or cost prohibitive to separate food material from packaging. A number of recent studies have demonstrated the potential benefits of copyrolysis of food wastes and other biomass materials with common plastics such as low- and high-density polyethylene (LDPE and HDPE): Serio et al. (2008), Önal et al., (2012), Abnisa and Daud (2014), Dewangan et al. (2016), Hassan et al. (2016), Yang et al. (2016), Tang et al. (2018), Uzoejinwa et al. (2018).

Another related thermochemical technology that is not as well developed at large scale is hydrothermal liquefaction (HTL), involving processing biomass at moderate temperature (280–370°C) and high pressure (10–25 MPa) to produce a crude-like bio-oil (Toor et al., 2011). This method is particularly well suited for wet feedstocks, because there is no need for drying. However, the corrosive operating conditions require the use of high-cost components that increase the capital investment. Despite the challenges encountered to date in developing HTL to commercial viability, it would appear that there is significant future potential because the technology can be applied to many different waste streams beyond food processing waste, including primary and secondary wastewater sludge, animal manure, fats, oils, and grease. Based on the recent study by Skaggs et al. (2018), combining all of these materials available in the conterminous United States as feedstock for HTL could potentially meet 23.9% of the national demand for aviation kerosene fuel.

3. Food waste biorefineries

As outlined above, there have been many laboratory-scale studies of the conversion of food processing wastes into renewable biofuels, including methane, hydrogen, biodiesel, ethanol, and butanol. However, effectively producing these fuels at commercial scale and achieving economic competiveness with incumbent technologies without subsidies has been challenging to say the least. In the United States, disposal rates for landfills and wastewater treatment are relatively low, and there is ample supply of low-cost natural gas. Therefore, without acquiring a credit for the production of "green" energy, there is little economic incentive to do so. With the current global focus on climate change and the expected future environmental impacts, it is possible that, in the future, greater value will be attributed to avoiding fossil carbon emissions, but this depends on many economic, political, and social factors that are difficult to predict.

One approach proposed to enhancing the economic performance of food waste-to-energy systems is the application of the so-called biorefinery concept, whereby one or more feedstocks are converted into a variety of value-added coproducts with very little residue, akin to a petroleum refinery. As stated by Cherubini (2010), "Biorefining is the sustainable processing of biomass into a spectrum of marketable products and energy." Although using mostly food waste is possible in principle (Carmona-Cabello et al., 2018; Dahiya et al., 2018), to make such systems practical, it is probably necessary to look beyond just the waste produced by the various stages of the food system (Fig. 4.1) to other resources that are potentially much larger and more consistent over time, such as lawn and forest residues, solid waste and wastewater from households, and even algae and seaweed. At the scale of a large, centralized facility, it is conceivable that a wide variety of specialty chemicals, bioproducts, fertilizers, solid, liquid, and gaseous biofuels could be produced, but such biorefinery outcomes can be achieved by combining a smaller number of subsystems around a single primary conversion system. For example, there has recently been growing interest in combining AD with thermochemical methods to valorize the solid fraction of the effluent from AD (also called "digestate") and also to potentially minimize the environmental impact of field spreading this stream (Opatakun et al., 2015; Peng and Pivato, 2017; Posmanik et al., 2017). Angenent et al. (2018) recommended enhancing the economic viability of AD by upgrading biogas into biomethane, converting carbon dioxide in biogas to more biomethane by hydrogenotrophic methanogenesis, generating cooling power from process heat, and producing bio-oil and a liquid biochemical product from organic matter. Perhaps the most compelling approach, albeit the farthest from commercial reality, would be to combine multiple technologies at a community scale where individual households, institutions, and businesses could extract value from the food and other organic wastes they generate, while maintaining investment and employment opportunities within the community itself (Fig. 4.4).

In developing potential biorefinery system architectures, it should be recognized that many opportunities exist for utilizing a single substrate (or associated coproducts) in more than one conversion system. For example, waste cooking oil (WCO) is a common precursor for biodiesel (via transesterification), but also has a very high biomethane potential of 641 L CH_4/kg (at STP), and thus is an excellent input for anaerobic codigestion (Labatut and Pronto, 2018). Apple processing waste (pomace) is an important food processing by-product generated worldwide and also in large quantities in our local region, because New York is the largest apple-producing state in the United States after Washington. This material has thus been widely studied in connection with various food waste-to-energy conversion processes, for example:

- Direct combustion of apple pomace can offset in-plant energy costs, but economic viability depends on waste flow rate, waste disposal cost, and fossil fuel price (Sargent et al., 1986).
- AD of apple slurry, waste, and pulp resulted in methane yields of 0.228–0.308 m^3/kg VS (Gunaseelan, 1997).
- Batch fermentation of apple pomace with *S. cerevisiae* at 5L working volume produced ethanol output of 0.4 g/g wet food waste (Parmar and Rupasinghe, 2013).

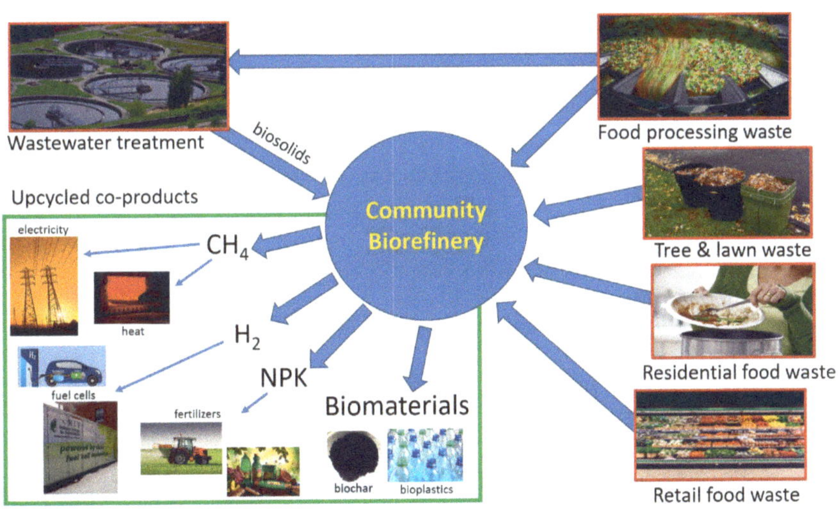

FIGURE 4.4

Conceptual community-scale biorefinery, with diverse materials (including food processing waste) combined to yield an array of value-added products.

- Apple pomace was used for butanol production with strains of *Clostridium acetobutylicum* and *Clostridium butylicum*. Yields of between 1.9% and 2.2% of fresh apple pomace were reported (Voget et al., 1985).
- Biohydrogen was produced via anaerobic fermentation of apple pomace with river sludge and achieved maximum cumulative yield of 101.08 mL/g total solid (TS) with an average H_2 production rate of 8.08 mL/g TS/h (Feng et al., 2010).
- Pyrolysis of apple pomace produced poor biochar yield but relatively high net energy output of coproducts of about 6 kJ/g of feedstock (Xu et al., 2011).

A similarly diverse collection of fundamental waste-to-energy studies exists for many common single-component food sector waste streams, including potato, coffee, tomato, etc., as well as mixed wastes from cafeterias, restaurants, etc.

One important consideration often overlooked in developing waste-to-energy systems is logistics. Even if ample feedstocks are available, demonstrated to be stable over time, and the core conversion technologies appear to be viable, significant resources will still be required to collect, characterize, monitor, handle, and transport both the influent and product materials. Additionally, one cannot underestimate the importance of the seamless flow of *information* in coordinating all these diverse operations and maintaining communications among the primary actors and stakeholders: waste generators, transporters/haulers, and waste-to-energy system operators, as well as local community leaders and policy makers (Armington et al., 2018). Regular dialogue among stakeholders enhances overall system efficiency and resilience and can lead to evolution of the portfolio of technologies being utilized over time.

4. Conclusions and future work

Food waste is a major global problem requiring immediate action to mitigate negative impacts on the environment. The food processing industry generates significant amounts of waste, but relative to other stages of the industrial food system produces by-product material that is more homogeneous and contaminant free and produced at high per-site rates. Many food processing operations have already established methods of using wastes as feedstock for value-added products such as nutraceuticals and other secondary food items, but there are opportunities for producing biofuels as another valorization strategy in cases where conventional disposal methods of landfilling, wastewater treatment, and incineration are currently being used.

AD, fermentation, TC, and transesterification can be used to produce a wide array of useful biofuels, including methane, hydrogen, ethanol, butanol, and biodiesel. The most economically favorable option depends upon many factors, with the specific phase of the waste material being perhaps the most important consideration. Whereas low moisture content solid wastes may be suitable for gasification or pyrolysis, liquid-phase wastes high in lipid and/or carbohydrate content are probably best suited for AD. A smaller subset of feedstocks appears to have viability for production of liquid fuels (ethanol and butanol) via fermentation processes. All of the technologies described in this chapter have been demonstrated at laboratory scale, and in most cases, the fundamental science is fairly well established. The challenge going forward is in applying sound engineering and design practices to minimize capital investment and operating costs and enhance adaptability to enable conversion systems to accept the widest possible array of diverse feedstocks. Because food waste resources are known to have significant spatial, temporal, and compositional variability, it is desirable for processes to be flexible enough to accept food and other more stable nonfood resources, such as municipal wastewater and biosolids, animal manure, and forest and lawn residue.

Although much of the research and development activities in food waste-to-energy processes have focused on large volumes of waste streams typically encountered in developed countries, there are many opportunities for utilizing available waste feedstocks in the rapidly expanding economies of Asia and Africa. For example, West Africa produces many commodity foodstuffs distributed worldwide and thus has massive resources of food processing wastes available for conversion to value-added products, including biofuels (Duku et al., 2011; Thomsen et al., 2014). A particularly compelling opportunity relates to cocoa bean production, amounting to a total of 4.7 million tons in 2016–17, where 75% of the whole fruit mass is the excess cocoa pod husk that is usually discarded (Lu et al., 2018).

Acknowledgments

The authors gratefully acknowledge Graduate Research Assistantship support for D. Rodríguez Alberto and partial support for T.A. Trabold provided by the US National Science Foundation (NSF) under Grant No. CBET-1639391.

References

Abnisa, F., Daud, W.M.A.W., 2014. A review on co-pyrolysis of biomass: an optional technique to obtain a high-grade pyrolysis oil. Energy Conversion and Management 87, 71–85.

Ahmed, I.I., Gupta, A.K., 2010. Pyrolysis and gasification of food waste: syngas characteristics and char gasification kinetics. Applied Energy 87 (1), 101–108.

Angenent, L.T., Usack, J.G., Xu, J., Hafenbradl, D., Posmanik, R., Tester, J.W., 2018. Integrating electrochemical, biological, physical, and thermochemical process units to expand the applicability of anaerobic digestion. Bioresource Technology 247, 1085–1094.

Arena, U., 2012. Process and technological aspects of municipal solid waste gasification. A review. Waste Management 32 (4), 625–639.

Armington, W.R., Chen, R.B., Babbitt, C.W., 2018. Challenges and innovations in food waste-to-energy management and logistics. In: Sustainable Food Waste-To-Energy Systems. Academic Press, pp. 259–271.

Arvanitoyannis, I.S., 2010. Waste Management for the Food Industries. Academic Press.

Carmona-Cabello, M., Garcia, I.L., Leiva-Candia, D., Dorado, M.P., 2018. Valorization of food waste based on its composition through the concept of biorefinery. Current Opinion in Green and Sustainable Chemistry 14, 67–79.

Chandrasekaran, M. (Ed.), 2013. Valorization of Food Processing By-Products. CRC Press.

Cherubini, F., 2010. The biorefinery concept: using biomass instead of oil for producing energy and chemicals. Energy Conversion and Management 51 (7), 1412–1421.

Dahiya, S., Kumar, A.N., Sravan, J.S., Chatterjee, S., Sarkar, O., Mohan, S.V., 2018. Food waste biorefinery: sustainable strategy for circular bioeconomy. Bioresource Technology 248, 2–12.

de Jong, W., van Ommen, J.R. (Eds.), 2014. Biomass as a Sustainable Energy Source for the Future: Fundamentals of Conversion Processes. John Wiley & Sons.

del Campo, A.G., Cañizares, P., Lobato, J., Rodrigo, M.A., Fernandez, F.J., 2012. Electricity production by integration of acidogenic fermentation of fruit juice wastewater and fuel cells. International Journal of Hydrogen Energy 37 (11), 9028–9037.

Dewangan, A., Pradhan, D., Singh, R.K., 2016. Co-pyrolysis of sugarcane bagasse and low-density polyethylene: influence of plastic on pyrolysis product yield. Fuel 185, 508–516.

Duku, M.H., Gu, S., Hagan, E.B., 2011. A comprehensive review of biomass resources and biofuels potential in Ghana. Renewable and Sustainable Energy Reviews 15 (1), 404–415.

Ebner, J.H., Rankin, M.J., Pronto, J., Labatut, R., Gooch, C., Williamson, A.A., Trabold, T.A., 2015. Greenhouse gas emissions analysis of a commercial-scale anaerobic co-digestion plant processing dairy manure and food waste. Environmental Science and Technology 49, 11199–11208.

Ebner, J.H., Labatut, R.A., Lodge, J.S., Williamson, A.A., Trabold, T.A., 2016. Anaerobic co-digestion of commercial food waste and dairy manure: characterizing biochemical parameters and synergistic effects. Waste Management 52, 286–294.

Feng, X., Wang, H., Wang, Y., Wang, X., Huang, J., 2010. Biohydrogen production from apple pomace by anaerobic fermentation with river sludge. International Journal of Hydrogen Energy 35 (7), 3058–3064.

Fiore, S., Ruffino, B., Campo, G., Roati, C., Zanetti, M.C., 2016. Scale-up evaluation of the anaerobic digestion of food-processing industrial wastes. Renewable Energy 96, 949–959.

Grycová, B., Koutník, I., Pryszcz, A., 2016. Pyrolysis process for the treatment of food waste. Bioresource Technology 218, 1203–1207.

Gunaseelan, V.N., 1997. Anaerobic digestion of biomass for methane production: a review. Biomass and Bioenergy 13 (1–2), 83–114.

Guran, S., 2018. Sustainable waste-to-energy technologies: gasification and pyrolysis. In: Sustainable Food Waste-To-Energy Systems. Academic Press, pp. 141–158.

Harvey, B.G., Meylemans, H.A., 2011. The role of butanol in the development of sustainable fuel technologies. Journal of Chemical Technology and Biotechnology 86 (1), 2–9.

Hassan, H., Lim, J.K., Hameed, B.H., 2016. Recent progress on biomass co-pyrolysis conversion into high-quality bio-oil. Bioresource Technology 221, 645–655.

Hegde, S., Trabold, T.A., 2018. Sustainable waste-to-energy technologies: fermentation. In: Sustainable Food Waste-To-Energy Systems. Academic Press, pp. 69–88.

Hegde, S., Lodge, J.S., Trabold, T.A., 2018. Characteristics of food processing wastes and their use in sustainable alcohol production. Renewable and Sustainable Energy Reviews 81, 510–523.

Knothe, G., Krahl, J., Van Gerpen, J. (Eds.), 2015. The Biodiesel Handbook. Elsevier.

Kosseva, M.R., Webb, C. (Eds.), 2013. Food Industry Wastes – Assessment and Recuperation of Commodities. Academic Press.

Labatut, R.A., Pronto, J.L., 2018. Sustainable waste-to-energy technologies: anaerobic digestion. In: Sustainable Food Waste-To-Energy Systems. Academic Press, pp. 47–67.

Lehmann, J., Joseph, S. (Eds.), 2015. Biochar for Environmental Management: Science, Technology and Implementation. Routledge.

Li, L., Peng, X., Wang, X., Wu, D., 2018. Anaerobic digestion of food waste: a review focusing on process stability. Bioresource Technology 248, 20–28.

Lin, C.S.K., Pfaltzgraff, L.A., Herrero-Davila, L., Mubofu, E.B., Abderrahim, S., Clark, J.H., Koutinas, A.A., Kopsahelis, N., Stamatelatou, K., Dickson, F., Thankappan, S., 2013. Food waste as a valuable resource for the production of chemicals, materials and fuels. Current situation and global perspective. Energy and Environmental Science 6 (2), 426–464.

Lipinski, B., Hanson, C., Lomax, J., Kitinoja, L., Waite, R., Searchinger, T., 2013. Reducing Food Loss and Waste. World Resources Institute Working Paper, pp. 1–40.

Liu, C., Li, H., Zhang, Y., Liu, C., 2016. Improve biogas production from low-organic-content sludge through high-solids anaerobic co-digestion with food waste. Bioresource Technology 219, 252–260.

Lu, F., Rodriguez-Garcia, J., Van Damme, I., Westwood, N., Shaw, L., Robinson, J.S., Warren, G., Chatzifragkou, A., Mason, S.M., Gomez, L., Faas, L., 2018. Valorisation strategies for cocoa pod husk and its fractions. Current Opinion in Green and Sustainable Chemistry 14, 80–88.

Łukajtis, R., Hołowacz, I., Kucharska, K., Glinka, M., Rybarczyk, P., Przyjazny, A., Kamiński, M., 2018. Hydrogen production from biomass using dark fermentation. Renewable and Sustainable Energy Reviews 91, 665–694.

National Research Council, 2015. A Framework for Assessing Effects of the Food System. National Academies Press.

Önal, E., Uzun, B.B., Pütün, A.E., 2012. An experimental study on bio-oil production from co-pyrolysis with potato skin and high-density polyethylene (HDPE). Fuel Processing Technology 104, 365–370.

Opatokun, S.A., Kan, T., Al Shoaibi, A., Srinivasakannan, C., Strezov, V., 2015. Characterization of food waste and its digestate as feedstock for thermochemical processing. Energy and Fuels 30 (3), 1589–1597.

Parmar, I., Rupasinghe, H.V., 2013. Bio-conversion of apple pomace into ethanol and acetic acid: enzymatic hydrolysis and fermentation. Bioresource Technology 130, 613–620.

Peng, W., Pivato, A., 2017. Sustainable management of digestate from the organic fraction of municipal solid waste and food waste under the concepts of back to earth alternatives and circular economy. Waste and Biomass Valorization 1–17.

Pham, T.P.T., Kaushik, R., Parshetti, G.K., Mahmood, R., Balasubramanian, R., 2015. Food waste-to-energy conversion technologies: current status and future directions. Waste Management 38, 399–408.

Posmanik, R., Labatut, R.A., Kim, A.H., Usack, J.G., Tester, J.W., Angenent, L.T., 2017. Coupling hydrothermal liquefaction and anaerobic digestion for energy valorization from model biomass feedstocks. Bioresource Technology 233, 134–143.

Rahman, S.N.A., Masdar, M.S., Rosli, M.I., Majlan, E.H., Husaini, T., 2015. Overview of biohydrogen production technologies and application in fuel cell. American Journal of Chemistry 5 (3A), 13–23.

ReFED, 2016. A Roadmap to Reduce US Food Waste by 20 Percent.

Ren, Y., Yu, M., Wu, C., Wang, Q., Gao, M., Huang, Q., Liu, Y., 2018. A comprehensive review on food waste anaerobic digestion: research updates and tendencies. Bioresource Technology 247, 1069–1076.

Sargent, S.A., Steffe, J.F., Pierson, T.R., 1986. The economic feasibility of in-plant combustion of apple processing wastes. Agricultural Wastes 15 (2), 85–96.

Serio, M., Kroo, E., Florczak, E., Wójtowicz, M., Wignarajah, K., Hogan, J., Fisher, J., 2008. Pyrolysis of Mixed Solid Food, Paper, and Packaging Wastes (No. 2008-01-2050). SAE Technical Paper.

Skaggs, R.L., Coleman, A.M., Seiple, T.E., Milbrandt, A.R., 2018. Waste-to-Energy biofuel production potential for selected feedstocks in the conterminous United States. Renewable and Sustainable Energy Reviews 82, 2640–2651.

Smith, M.D., Rabbitt, M.P., Coleman-Jensen, A., 2017. Who are the world's food insecure? New evidence from the Food and Agriculture Organization's food insecurity experience scale. World Development 93, 402–412.

Sørum, L., Grønli, M.G., Hustad, J.E., 2001. Pyrolysis characteristics and kinetics of municipal solid wastes. Fuel 80 (9), 1217–1227.

Tang, Y., Huang, Q., Sun, K., Chi, Y., Yan, J., 2018. Co-pyrolysis characteristics and kinetic analysis of organic food waste and plastic. Bioresource Technology 249, 16–23.

Thomsen, S.T., Kádár, Z., Schmidt, J.E., 2014. Compositional analysis and projected biofuel potentials from common West African agricultural residues. Biomass and Bioenergy 63, 210–217.

Toor, S.S., Rosendahl, L., Rudolf, A., 2011. Hydrothermal liquefaction of biomass: a review of subcritical water technologies. Energy 36 (5), 2328–2342.

Trabold, T.A., Babbitt, C.W. (Eds.), 2018. Sustainable Food Waste-to-Energy Systems. Academic Press.

Trabold, T.A., Win, S.S., Hegde, S., 2018. Waste resources in the food supply chain. In: Sustainable Food Waste-to-Energy Systems. Academic Press, pp. 11–28.

Usack, J.G., Van Doren, L.G., Posmanik, R., Labatut, R.A., Tester, J.W., Angenent, L.T., 2018. An evaluation of anaerobic co-digestion implementation on New York State dairy farms using an environmental and economic life-cycle framework. Applied Energy 211, 28–40.

Uzoejinwa, B.B., He, X., Wang, S., Abomohra, A.E.F., Hu, Y., Wang, Q., 2018. Co-pyrolysis of biomass and waste plastics as a thermochemical conversion technology for high-grade biofuel production: recent progress and future directions elsewhere worldwide. Energy Conversion and Management 163, 468–492.

Van Gerpen, J., 2005. Biodiesel processing and production. Fuel Processing Technology 86 (10), 1097–1107.

Venderbosch, R.H., Prins, W., 2010. Fast pyrolysis technology development. Biofuels, Bioproducts and Biorefining 4 (2), 178–208.

Voget, C.E., Mignone, C.F., Ertola, R.J., 1985. Butanol production from apple pomace. Biotechnology Letters 7 (1), 43–46.

Xu, R., Ferrante, L., Hall, K., Briens, C., Berruti, F., 2011. Thermal self-sustainability of biochar production by pyrolysis. Journal of Analytical and Applied Pyrolysis 91 (1), 55–66.

Xu, F., Li, Y., Ge, X., Yang, L., Li, Y., 2018. Anaerobic digestion of food waste–Challenges and opportunities. Bioresource Technology 247, 1047–1058.

Yang, J., Rizkiana, J., Widayatno, W.B., Karnjanakom, S., Kaewpanha, M., Hao, X., Abudula, A., Guan, G., 2016. Fast co-pyrolysis of low density polyethylene and biomass residue for oil production. Energy Conversion and Management 120, 422–429.

Yasin, N.H.M., Mumtaz, T., Hassan, M.A., 2013. Food waste and food processing waste for biohydrogen production: a review. Journal of Environmental Management 130, 375–385.

Zevenhoven-Onderwater, M., Backman, R., Skrifvars, B.-J., Hupa, M., 2001. The ash chemistry in fluidised bed gasification of biomass fuels. Part I: predicting the chemistry of melting ashes and ash-bed material interaction. Fuel 80, 1489–1502.

CHAPTER

Food industry by-products valorization and new ingredients: Cases of study

5

Antonio D. Rodriguez-Lopez[1], Bruno Melgar[2], Carmela Conidi[3], Lillian Barros[4], Isabel C.F.R. Ferreira[4], Alfredo Cassano[3], Esperanza M. Garcia-Castello[2]

[1]*Institute for Industrial, Radiophysical and Environmental Safety, ISIRYM, Universitat Politècnica de València, Valencia, Spain;* [2]*Institute of Food Engineering for Development, IuIAD, Universitat Politècnica de València, Valencia, Spain;* [3]*Institute on Membrane Technology, ITM-CNR, Rende, Cosenza, Italy;* [4]*Centro de Investigação de Montanha, CIMO, Instituto Politécnico de Bragança, Campus de Santa Apolónia, Bragança, Portugal*

1. Introduction: waste and wastewaters from food industry

Along the time, different institutions have promoted and attempted to define what can be considered as food waste. In this sense, in Europe as well as in the United States, it was established a similar waste hierarchy for the waste treatment: (1) reduction; (2) reuse; (3) material recovery; (4) energy recovery; and (5) waste disposal (Directive, 2008/98/EC; USEPA, 2001).

In 2009, the Food and Agriculture Organization (FAO) of the United Nations published a report that estimated the world population will reach in 2050 around 9.1 billion and the main increment of population will take place in developing countries (FAO, 2009). Besides, the FAO also performed a study published in 2011 that resulted in the conclusion that one-third of food produced all over the world gets lost during the food chain from farm (agricultural production) to fork (consumption in households) (FAO, 2011; Carciochi et al., 2017). According to the FAO (2011), if food lost occurs during the production, postharvest, and processing stages, then it is called food losses. On the other hand, when food lost is in the retail or consumption stages, it is named food wastes.

In 2012, the ISEKI Food Association established the term "food waste recovery" with the meaning of "extraction of valuable compounds from waste by-products" (Galanakis, 2018; ISEKI, 2012).

The European Commission, in 2014, redefined the concept "food waste" as the "loss of food, including nonedible parts, from the food supply chain, that are not sent for animal feed, redistribution ..." (Galanakis, 2018). Considering the above discussion, here and after, this meaning for food waste will be used.

Due to the general characteristics of food wastes (organic load, high moisture, and microbial activity, among others), they generate different problems on the

Sustainability of the Food System. https://doi.org/10.1016/B978-0-12-818293-2.00005-7

environment such as CO_2 emissions, toxicity in surface and ground waters, changes in soil quality, etc. (Nayak and Bhushan, 2019; FAO, 2011).

Reduction of food waste is one of the goals of the United Nations to achieve a more sustainable world by 2030 (United Nations, 2015). According to this idea, the European Union has proposed different strategies to promote a reduction in waste production, facilitating the food donation and improving labeling. On the other hand, to reduce the amounts of wastes, it is possible also to focus the efforts on the recovery of substances and compounds included in waste streams from food industry (Galanakis, 2018).

The environmentally sustainable food production has to reconsider wastes as by-products that can be transformed to provide valuable compounds (antioxidants, fiber, and fuels, among others) and then be used as new products or raw materials in the food industry or even applied in other sectors such as pharmaceutical, polymer, and energy industries (Carciochi et al., 2017).

In this sense, the term biorefinery has been appeared as the whole strategy focused on a green economy, considering all steps in the food chain production, for the achievement of a zero-waste concept (Fava et al., 2015; Nayak and Bhushan, 2019; Carciochi et al., 2017). In this context, biocompounds are considered as biological compounds that can be used to increase the nutrient content in human diets and are extracted mainly from plant materials such as fungi, legumes, cereals, grains, vegetables, and fruits (Radrigán et al., 2017; Quintin et al., 2019).

In addition, due to the increasing consumers' request on natural ingredients derived from natural sources with almost no content in toxins, pesticides, etc., and the growing demand on nutraceutical products, some food industries are including natural food products, as wild products, as a new source of healthy food resources.

2. Food ingredients obtaining

The extraction, recovery, concentration, and purification operations of valuable ingredients (antioxidants, vitamins, proteins, sugars, colorants, etc.) from food waste and from nonwidely exploited harvest such as wild fruits become a key step to guarantee both environmental and economic sustainability of the whole obtaining process. Furthermore, whatever are the steps considered in the ingredients obtaining, it is necessary to maintain the stability of the desired healthy properties of these ingredients.

All of these operations are based mainly on the application of a combination of chemical, thermal, and physical processes on food or their wastes matrices (Nayak and Bhushan, 2019). There are three fundamental issues to consider: (1) the concentration of the ingredients to be extracted is usually low; consequently, the cost of the extraction and concentration can be high; (2) in general, target ingredients are thermolabile and this limit the use of some technologies (Galanakis, 2018; Carciochi et al., 2017); and (3) it is necessary to evaluate the toxicity and amount of residues obtained by using some extraction operations.

Available technologies used for the extraction and recovery of biocompounds can be classified by conventional and nonconventional technologies. Conventional technologies are considered as classic and traditional operations (solvent extraction, ultrafiltration [UF], microfiltration [MF], nanofiltration [NF], adsorption, centrifugation, alcohol precipitation, electrodialysis, thermal and/or vacuum concentration, among others), whereas nonconventional are those technologies that are more recently studied and, in some cases, implemented in industry. Some examples are ultrasound-assisted extraction (UAE), microwave-assisted extraction (MAE), liquid membranes, low-temperature plasma treatment, colloidal gas aphrons, enzyme-assisted extraction, nanotechnology, and pulsed electric field (Galanakis, 2012; Carciochi et al., 2017; Radrigán et al., 2017; Nayak and Bhushan, 2019; Renard, 2018).

Some of these technologies and examples of application are summarized in Table 5.1 according to their widespread use in food industry or their promising applicability and focused on the environmental and health sustainability.

3. Successful study cases

3.1 Wild vegetables as a source of value-added compounds

One of the main interests in food by-products rises from different axis like recovery and exploitation of natural alternative resources, pollution reduction, environmental awareness, and many others that joint in conjunction to accomplish food security (Galanakis, 2012). Worldwide malnutrition and food inequity are important subjects that have led researchers to investigate alternatives products to address these problematics.

Unbalanced diets and food scarcity are the main problems that converge in undernourished populations; therefore, for a diet to be balanced, it must be composed of a wide range of food groups that ensure the delivery of nutrients that the body requires (Roberfroid, 2011). Even though fruits and vegetables are relatively accessible for some social groups, in certain regions of the world, some factors such as adverse environmental conditions, overpopulation, social-political problems, and poverty affect the access to agroindustrial crops and the consequent famine and malnutrition, on those regions. However, although a large percentage of our diet relays on technified agriculture, wild fruits, vegetables, and spices are still being used in some social groups as a source of nutrition and folk medicine, due to their high content in phytochemicals (Bvenura & Sivakumar, 2017).

In this sense, the interest for alternative vegetables, which present benefits and unique qualities, has grown in recent years (Martins and Ferreira, 2017), and therefore, they are described in this section. In this section, some vegetables rich in phenolic compounds have been selected.

Table 5.1 Characteristics and extraction examples of different conventional and nonconventional extracting operations.

	Characteristics	Extraction examples
Conventional treatments		
Solvent extraction	Characterized by the use of organic solvents, mainly ethanol and high time and energy consumption. Solvents have to be considered as GRAS (generally recognized as safe) by USFDA (1–2).	Pectin from orange peel and grapefruit peel (3–4). Phenolics from pomegranate wastes (5).
Membrane technology	Used for the treatment of liquid streams from food industry. Based on a tangential filtration in which a semipermeable membrane is used as a separation agent (6). Membrane processes most used are microfiltration, ultrafiltration, nanofiltration, and reverse osmosis. Other membrane operations are membrane distillation and osmotic distillation.	Phenolics from olive mill wastewaters and artichoke processing wastewaters (7–9). Betalains and colorants from different sources (10).
Nonconventional treatments		
Ultrasound-assisted extraction (UAE)	Alternative to conventional extraction technologies (11–12). Based on the use of frequencies between 20 kHz and 100 MHz (13–14) that generate the cavitation effect and facilitate the extraction of ingredients from the solid matrix (15). UAE promotes a reduction of extracting time and solvent amounts and increases the quality and concentration of biocompounds in extracts.	Anthocyanins from carrot pomace (16). Flavonoids from grapefruit peels and kiwifruit juice (17–18). Carbohydrates from herb, wood, algae, and mushrooms (15).
Microwave-assisted extraction (MAE)	Based on the use of high frequencies (300 MHz–300 GHz). Reduces the extraction time since it promotes a quicker heating, decreases thermal gradients, and minimizes the need of high amounts of extracting solvents (13, 19).	Polyphenols from pomegranate peels and grape marcs (20–21). Lycopene from tomato peels (22). Pectins from orange and mango peels (23–24).

Table 5.1 Characteristics and extraction examples of different conventional and nonconventional extracting operations.—*cont'd*

	Characteristics	Extraction examples
Supercritical fluid extraction	Based on the use of supercritical fluid (SF) characterized by its dissolving properties of a liquid and a viscosity and diffusivity close to vapors. SF reduces energy consumption and organic solvents resulting in a higher quality of extracts (11, 13, 25–28). The most SF used is CO_2.	Antioxidants and polyphenols from apple pomace, agroindustrial soybean wastes and hazelnut, coffee, and grape wastes (29–30).
Pulsed electric field	Based on the application of high-voltage pulses to a product using electrodes. High-voltage pulses provoke cell membrane rupture, a decreasing of the extraction time, and an increment of the permeability of cell membranes (11, 13, 25–27)	Proteins from waste chicken meat (32). Pigments as carotenoids from tomatoes (33).
Enzyme-assisted extraction (EAE)	The EAE is getting more importance since it is considered a green extraction method due to the organic and natural nature of enzymes. It allows the rupture of cell membranes and favors the recovery of biocompounds (13, 25 −27).	Proteins and antioxidants from sesame bran (34).

1. Cvjetko Bubalo et al. (2018); 2. USFDA (2016); 3. Georgiev et al. (2012); 4. Bagherian et al. (2011); 5. Saffarzadeh-Matin and Khosrowshahi (2017); 6. Galanakis et al. (2016); 7. Garcia-Castello et al., 2010; 8. Conidi et al. (2014); 9. Conidi et al. (2015); 10. Raghavarao et al. (2014); 11. Galanakis (2012); 12. Paniwnyk et al. (2017); 13. Radrigán et al., 2017; 14. Lavilla and Bendicho (2017); 15. Herrera and Luque de Castro (2005); 16. Agcam, Akyıldız and Balasubramaniam (2017); 17. Garcia-Castello et al., 2015; 18. Wang et al. (2019); 19. Martín and Navarrete (2018); 20. Kaderides et al. (2019); 21. Garrido et al. (2019); 22. Eh and Teoh (2012); 23. Maran et al. (2013); 24. Maran et al. (2015); 25. Carciochi et al., 2017; 26. Renard, 2018; 27. Quintin et al., 2019; 28. Galanakis (2018); 29. Ferrentino et al. (2018); 30. Alvarez et al. (2019); 31. Manna et al. (2015); 32. Ghosh et al. (2019); 33. Bot el al., 2018; 34. Görgüç et al. (2019).

Besides the plants mentioned in Table 5.2, there are many other wild vegetables with a great potential of study due to their bioactive profile. Scarcity of precise identification of these compounds has only shown in some documents general screening of the phenolic profiling, but they have not been studied in depth, leaving a great potential for the research of recovery and exploitation of more natural phytochemicals that could be employed as new functional ingredients (Lin and Lin, 1997; El-Kader et al., 2006).

Table 5.2 Selected potential wild vegetables rich in bioactive compounds.

Scientific name	Common name	Uses	Distribution	Principal bioactive compounds	References
Amaranthus spinosus	Quelite/phak khom	Food, beverage	America, asia	Amaranthine, quercetin, kaempferol, rutin	1,2,3
Apium nodiflorum (L) Lag.	Erba canella/ rabaças	Food, ethnomedicine	Italy	Germacrene D, allo-ocimene, limonene	4,5
Calendula spp.	Calendula/ margarita	Food, cosmetics, ethnomedicine	Mediterranean	Caffeic acid, quercetin	6,7
Cichorium intybus L.	Belgian endive/ chicory	Food, ethnomedicine	Europe*	Chicoric acid, caffeoylquinic acid, quercetin	8,9,10
Foeniculum vulgare Mill.	Fennel	Food, ethnomedicine	Spread around the world	Caffeoylquinic acid, quercetin, kaempferol, isorhamnetin	11,12,13
Portulaca oleracea L.	Verdolaga/ purslane	Food, ethnomedicine	Spread around the world	Kaempferol, apigenin, luteolin, myricetin	14
Carduus tenuiflorus Curtis	Sheep thistle	Ethnomedicine	Spread around the world	n/f	15
Beta macrocarpa Guss.	Acelga de fruto grande	Ethnomedicine	Mediterranean	n/f	16

*1. Balakrishnan et al. (2011); 2. Hilou et al. (2006); 3. Peter and Gandhi (2017); 4. Maxia et al. (2012); 5. Carvalho et al. (2011); 6. Faustino et al. (2018); 7. Guijarro-Real et al. (2019); 8. Abbas et al. (2014); 9. Gallucci et al. (2017); 10. Tardugno et al. (2018); 11. Caleja et al. (2014); 12. Levorato et al. (2018); 13. Qurishi et al. (2012); 14. Farkhondeh and Samarghandian (2019); 15. Enrique de Briano, Acciaresi and Briano (2013); 16. Iamonico (2019); *Majoritarian Italian usage.*

3.2 Citric industry: traditional processing and new environmental and integrated processes for the recovery of food ingredients

The citrus processing industry plays an important role in the agroindustrial sector. According to data published by the FAO (2016), citrus represents the third most important fruit crop with a production of about 124,246,000 tons.

In contrast with other types of fruit, citrus fruits can be consumed mostly fresh or pressed to produce juice. Furthermore, citrus fruits can be processed to obtain other food products such as dehydrated citrus products such as jams, jellies, or marmalades.

Citrus by-products are considered to be an economic and renewable source of valuable compounds such as minerals, dietary fibers, oils, lipids, pectins, and bioactive compounds, such as polyphenols and carotenoids.

To obtain them, it is possible to use conventional techniques, but recently, it has been developed green or clean techniques based on reduced use of energy, short extraction time, decrease of solvent consumption, overall enhancement of extraction rate, enhancement of the quality extracts, improvement of aqueous extraction processes, and improved extraction of heat-sensitive compounds (Vilkhu et al., 2008).

Different studies have analyzed the use of green technologies for the recovery of bioactive compounds from citrus by-products. MAE, UAE, and membrane processes are among the most innovative techniques developed recently (Conidi et al., 2018; Khan et al., 2010).

An optimized MAE method for the extraction of polyphenol compounds from citrus mandarin juice peels has been reported by Hayat et al. (2009), and higher extraction yields of phenolic acids and antioxidant capacity of the extracts were obtained, if compared with traditional technology.

Garcia-Castello et al. (2015) compared conventional extraction and UAE of flavonoids from grapefruit peels, and the experimental results showed better performance of UAE in terms of extraction yields (on average total phenolic content 50% and total antioxidant activity [TAA] 66% higher), at lower temperature and extraction time, when compared with conventional solvent extraction.

Membrane processes have been also successfully investigated for the recovery of biological active compounds from citrus by-products, thus transforming these by-products to source material for high added-value compounds.

Conidi et al. (2012) investigated the use of spiral-wound NF membranes with different molecular weight cutoff (MWCO) (from 250 to 1000 Da) and polymeric material (polyamide, polypiperazine amide, and polyethersulfone) for the separation and concentration of phenolic compounds from red range press liquors.

In a previous work, the same authors studied the concentration of polyphenols from bergamot juice, a by-product of the essential oil production, using tight UF and NF membranes (Conidi et al., 2011).

As a green extraction technique, an innovative process on orange peel wastes was developed by Boukroufa et al. (2015). The process involved the use of MAE and

UAE in an integrated system, for the recovery of essential oil, polyphenols, and pectin from orange peel.

In a similar approach, Fidalgo et al. (2016) studied the extraction of pectin and D-limonene from waste orange and lemon peel by an innovative eco-friendly process using water as dispersing medium and microwaves as energy source. The investigated process permitted to obtain high yields of extracted selected compounds, with high quality and environmental viability.

A sustainable process based on the integration of UF, NF, and osmotic distillation (OD) membrane processes for the recovery and concentration of flavonoids from orange press liquor was investigated by Cassano et al. (2014). The press liquor was previously clarified by a UF membrane module in hollow fiber configuration and concentrated from 10 to 32°Brix by using an NF membrane in spiral-wound configuration. The NF retentate was finally concentrated by OD up to 47°Brix, working in conditions of low mechanical and thermal damage. The concentrated extracts, produced at low temperatures without thermal damage to the compounds of interest, offered interesting perspectives for the use of these products as natural colorants and/or for nutraceutical applications.

A schematic of integrated green processes for the recovery of valuable compounds from citrus by-products is illustrated in Fig. 5.1.

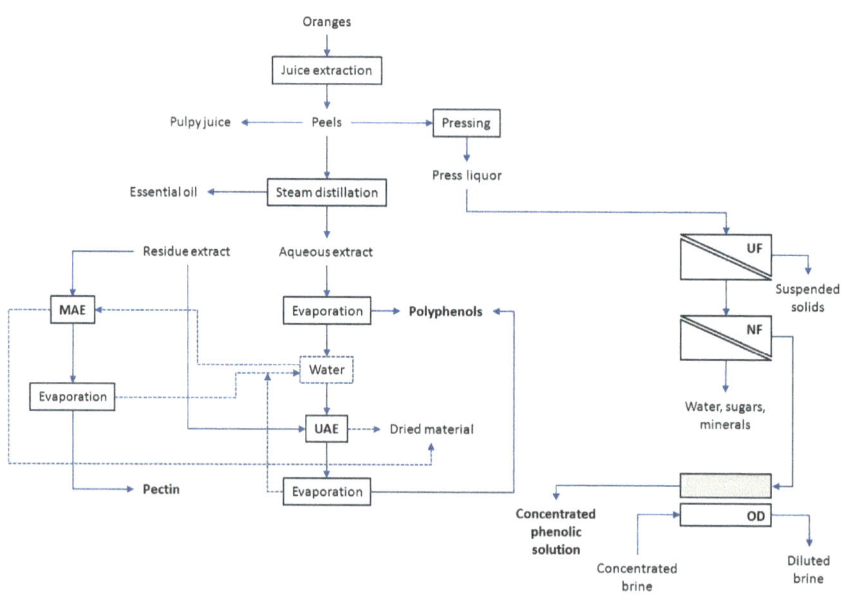

FIGURE 5.1

Integrated green processes for the recovery of valuable compounds from citrus by-products (*MAE*, Microwave-assisted extraction; *UAE*, Ultrasound-assisted extraction; *UF*, Ultrafiltration; *NF*, Nanofiltration; *OD*, Osmotic distillation).

Adapted from Boukroufa et al., 2015 and Cassano et al., 2014.

3.3 Olive oil industry: traditional processing and new environmental and integrated processes for the recovery of food ingredients

Olive oil production is one of the most traditional agricultural industries with a great economic importance in most of the Mediterranean countries.

The by-products obtained are a solid fraction (called olive pomace), and a liquid fraction called olive mill wastewater (OMW). This liquid stream is the mix between olive vegetation water and the possible underprocess added water.

The organic fraction of OMWs contains sugars, tannins, polyphenols, polyalcohols, pectins, lipids, proteins, and organic acids (Cassano et al., 2011). OMWs are considered highly polluting effluents due to their high organic load and require a complex management and disposal system due to their negative effects on the environmental (such as antimicrobial, ecotoxic, and phytotoxic properties) (Paraskeva et al., 2007).

Different management options have been proposed for the treatment of olive wastes, but independently by the efficiency of those physical, chemical, and biological treatments, the disposal of great amounts of sludge produced remains a significant problem in the OMW treatment.

According to hierarchy of the EU and EPA aims, aforementioned in the Section 1 of this chapter, it is increasing attention to the valorization of these streams through the recovery and/or the biotransformation of their organic matter as the recovery of antioxidant substances and phenols, which can be exploited in pharmaceutical, food, and cosmetical applications (Kalogerakis et al., 2013). For instance, their introduction in foodstuffs could decrease the use of synthetic antioxidants, such as butylate hydroxyanisole, butylated hydroxytoluene (BTH), and tert-butylhydroquinone. Due to their potential negative effects on health, these synthetic antioxidants are restricted to maximum concentrations in foodstuffs (Araújo et al., 2015).

The efficient recovery of phenolic compounds has been extensively studied and optimized to maximize their reintroduction in the food chain and contribute to a higher valorization and better management of OMWs (Kalogerakis et al., 2013).

Excellent biological properties in terms of antioxidant, free radical scavenging, antimicrobial, and anticarcinogenic activities of the polyphenols of OMWs have been reviewed by different authors (Obied et al., 2005; Araújo et al., 2015). The most important biophenols include benzoic acid and hydroxycinnamic acid derivatives and, in larger amounts, tyrosol, hydroxytyrosol, and oleuropein.

The efficiency of a green approach for the recovery of phenols from OMWs based on the use of UAE was studied by Klen and Vodopivec (2011). Results showed higher recoveries at both levels of individual and total phenol yields in comparison with conventional technologies (filtration, solid-phase, and liquid-liquid extraction).

Membrane processes have been also successfully used as "green processes" for the recovery of water, organic compounds, and polyphenols from OMWs. These processes, mostly in a sequential form or combined with other separation technologies, successfully meet the requirement for the recovery, purification, and concentration of antioxidants from OMWs suitable for food or pharmaceutical formulations (Conidi et al., 2018).

Garcia-Castello et al. (2010) investigated the recovery and concentration of OMW phenolic compounds from OMWs using an integrated membrane system based on conventional pressure-driven processes (MF and NF) and OD. Results showed that about 78% of the initial phenolic compound content was recovered in the microfiltered permeate solution. The MF permeate was then submitted to NF, after which a concentrated solution containing approximately 0.5 g/L phenolic compounds, with hydroxytyrosol representing 56% of the total, was obtained by processing the NF permeate with OD.

A combination of UF and NF processes for the recovery of biologically active compounds and water from OMWs was investigated by Cassano et al. (2013). In this approach, raw wastewaters were firstly clarified with a UF hollow fiber membrane module (HFS, Toray) to remove suspended solids and colloidal substances. Afterward, the UF permeate was processed by a UF unit equipped with a flat-sheet membrane. The produced permeate stream was finally concentrated by using an NF unit equipped with a spiral-wound NF membrane. UF membranes allowed to recover most part of phenolic compounds in the permeate streams due to their low rejection toward these components (in the range of 26%–31%, respectively). On the other hand, a rejection higher than 93% toward total polyphenols was measured by the NF90 membrane. Free low-molecular-weight polyphenols, including caffeic acid, p-coumaric, catechol, tyrosol, procatechuic acid, and hydroxyl-tyrosol, were completely retained, in accordance with the estimated MWCO (200 Da) of the membrane and the MW of target compounds (138–284 g/mol). The NF retentate, enriched in polyphenols, was considered suitable for food, cosmetic, and pharmaceutical formulations, whereas the purified aqueous fraction (NF permeate) could be reused in the olive oil extraction or for membrane cleaning.

Conidi et al. (2014) evaluated the performance of a multiphase biocatalytic membrane reactor in the conversion of a UF permeate containing oleuropein into oleuropein aglycon catalyzed by β-glucosidase immobilized in a polymeric membrane. The maximum oleuropein conversion reached was about 45.7%, and the reaction rate was about 2×10^{-4} mmol/min cm^3.

An integrated membrane system based on a sequential combination of pressure-driven membrane operations and innovative membrane systems has been recently implemented by Bazzarelli et al. (2016). OMWs were firstly acidified with sulfuric acid to achieve a complete removal of suspended solids and then clarified by MF. The MF permeate was treated by NF to obtain water from the permeate side and a concentrated polyphenolic solution from the retentate side. The NF retentate was dewatered by OD by using a polypropylene hollow fiber membrane contactor (Liqui-Cell Extra-Flow 2.5 × 8″, Membrana) and calcium chloride dihydrate at 60 w/w% as stripping solution. The OD retentate was finally encapsulated in a water-in-oil emulsion by membrane emulsification. According to the process mass balance, about 1.5 kg of phenolic compounds (85% of the initial phenolic content) and 800 L of purified water can be recovered from 1 m^3 of OMW (Fig. 5.2).

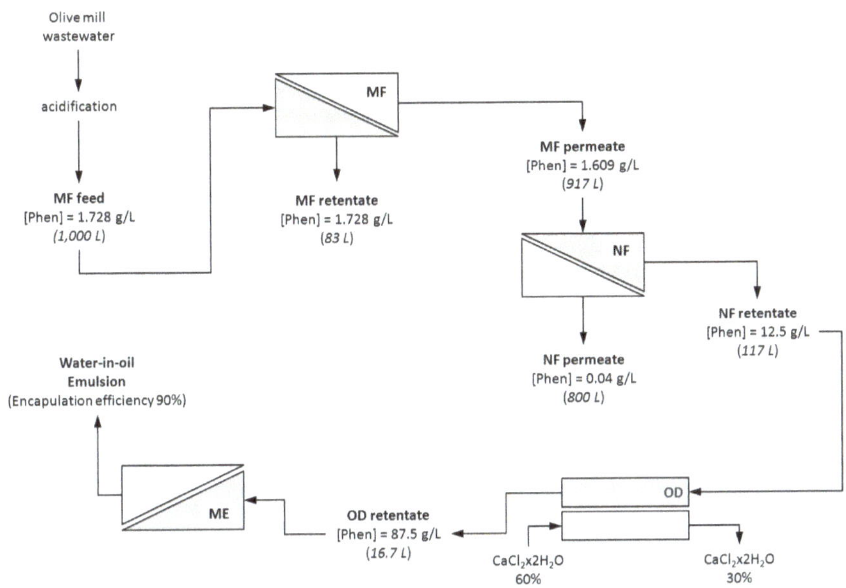

FIGURE 5.2

Integrated membrane process used for the fractionation of olive mill wastewaters (*MF*, Microfiltration; *NF*, Nanofiltration; *OD*, Osmotic distillation; *ME*, Membrane emulsification).

Adapted from Bazzarelli, F., Piacentini, E., Poerio, T., Mazzei, R., Cassano, A. & Giorno, L. (2016). Advances in membrane operations for water purification and biophenols recovery/valorization from OMWWs. Journal of Membrane Science, 497, 402–409. https://doi.org/10.1016/j.memsci.2015.09.049.

3.4 Cactus pear industry

3.4.1 Characterization and evaluation of new ingredients from fruits and their wastes

Within the Cactaceae family, *Opuntia* genus has strong cultural and economic significance throughout the Americas and some arid environments regions in the world as crops for their alimentary and forage products. World scientific community has constantly increased their publications, with 249 records including the words "*Opuntia* fruit" in their titles, until January 2019 (WOS, 2019). Scientific records mainly come from Italy, Mexico, United States, and Morocco with more than 50% of the total publications. Among these, the variety *ficus-indica* (OFI), commonly known as prickly pear or cactus pear, is widely studied, although some authors have also researched on some of the 300 wild more uncommon species (spp.) (Morales-Martinez et al., 2018; Morales et al., 2014; Melgar et al., 2017).

Opuntia spp. fruits vary substantially on their morphological and chemical attributes, for instance, the percentage of peel fluctuates from 25% in OFI to nearly 65% in *Opuntia engelmannii* variety (Melgar et al., 2017). Since only the edible part of the fruit is used in the food chains, a huge amount of waste could be transformed into a potentially useful by-product, providing high added-value compounds with food, cosmetic, and pharmaceutical applications (Galanakis, 2012).

Table 5.3 presents a wide variety of bioactive compounds recovered from *Opuntia* spp. by-products. Constituents such as pectins, phenolic acids, flavonoids, carotenoids, betalains, tocopherols, and phytosterols have been previously described (Amaya-Cruz et al., 2019; Morales-Martínez et al., 2018; Morales et al., 2014; Melgar et al., 2017). These compounds could provide ample possibilities of use of this by-product as a source of new ingredients with added healthy properties (Valdés and León, 2017).

Despite the importance of phenolic and nonphenolic compounds within *Opuntia* spp. by-products, this section focuses on their inner natural pigments. Betalains are water-soluble nitrogen-containing compounds present in a restricted number of families of the plant order of Caryophyllales. Besides anthocyanins, carotenoids, and chlorophylls, betalains complement the natural widespread molecules used for coloring purposes (Delgado-Vargas and Paredes-Lopez, 2003). These compounds are situated in the vacuoles of the plant and their basic structure consisting of a moiety of betalamic acid and, depending on the residue, could be classified as red/purple betacyanins and yellow-orange betaxanthins (when cyclo-Dopa [cyclo-3-(3,4-dihydroxyphenylalanine)] and hydroxycinnamic acid derivatives or sugars, and amines or amino acids residues also bind to betalamic acid, respectively) (Herbach et al., 2006).

The ever-increasing demand for new natural ingredients has brought the development of a plethora of pigment extraction systems and their description in the scientific literature. Sample preparation is considered a crucial part of this process and frequently, eco-friendly solvents and technologies, which contemplate recovery, recycling, and sustainability, are the most commonly used (Płotka-Wasylka et al., 2017). Otherwise, generation of hazardous toxic waste could result contradictory to the concept of by-products recovery. For instance, due to the polarity of betalains and morphological properties of the by-products, solid-liquid extractions are the most commonly used (Barba et al., 2017), with aqueous-based solvents used mainly as the extracting phases, and also often mixed with other volatile organic compounds applied as organic solvents (Amaya-Cruz et al., 2019; Koubaa et al., 2016; Mena et al., 2018; Morales et al., 2015; Melgar et al., 2017). Extraction techniques and characterization of the complex nonhomogeneous extracts are also varied and differed among the scientific community. In Table 5.4, frequently used solvents and techniques are listed along with recurrent final detection techniques.

Besides betalains' ability of pigmentation, their antioxidant ability and capacity to absorb free radical makes this natural colorant an excellent alternative to the artificial counterpart, due to their potential benefits against chronical diseases such as cardiovascular, inflammatory, diabetes, cancer, oxidative stress, and other diseases associated with aging (Amaya-Cruz et al., 2019; Garcia-Cruz et al., 2017; Herbach et al., 2006; Sawicki et al., 2016; Stintzing et al., 2005).

Table 5.3 Bioactive compounds cited in *Opuntia* spp. peels.

Family	Compound name	Variety	Author
Hydroxybenzoic acids	2-Hydroxybenzoic acid	OFI	Amaya-Cruz et al. (2019)
	2,3-Dihydroxybenzoic acid	OFI	Amaya-Cruz et al. (2019)
	2,4-Dihydroxybenzoic acid	OFI	Amaya-Cruz et al. (2019)
	3-Hydroxybenzoic acid	OFI	Amaya-Cruz et al. (2019)
	3,5-Dihydroxybenzoic acid	OFI	Amaya-Cruz et al. (2019)
	4-Hydroxybenzoic acid	OFI; OM; OMT	Amaya-Cruz et al. (2019), Ndhlala et al. (2007), Guzmán-Maldonado et al. (2010)
	4-Hydroxybenzoic acid 4-*O*-glucoside	OFI	Amaya-Cruz et al. (2019)
	Benzoic acid	OFI	Amaya-Cruz et al. (2019)
	Ellagic acid	OFI	Amaya-Cruz et al. (2019)
	Ellagic acid acetyl-arabinoside	OFI	Amaya-Cruz et al. (2019)
	Ellagic acid arabinoside	OFI	Amaya-Cruz et al. (2019)
	Eucomic acid	OFI	Melgar et al. (2017), Chougui et al. (2015)
	Gallic acid	OMT	Guzmán-Maldonado et al. (2010)
	Gallic acid 3-*O*-gallate	OFI	Amaya-Cruz et al. (2019)
	Isovanillic acid	OFI	Amaya-Cruz et al. (2019)
	Piscidic acid	OFI	Melgar et al. (2017), Chougui et al. (2015)
	Protocatechuic acid	OFI; OM	Amaya-Cruz et al. (2019), Ndhlala et al. (2007), Mena et al. (2018)
	Protocatechuic acid 4-*O*-glucoside	OFI	Amaya-Cruz et al. (2019)
	Syringic acid	OFI	Amaya-Cruz et al. (2019)
	Vanillic acid	OFI; OMT	Amaya-Cruz et al. (2019), Guzmán-Maldonado et al. (2010)

Continued

Table 5.3 Bioactive compounds cited in *Opuntia* spp. peels.—*cont'd*

Family	Compound name	Variety	Author
Hydroxycinnamic acids	3-*O*-Caffeoylquinic acid	OFI	Amaya-Cruz et al. (2019)
	3-*p*-Coumaroylquinic acid	OFI	Amaya-Cruz et al. (2019)
	3,4-*O*-Dicaffeoylquinic acid	OFI	Amaya-Cruz et al. (2019)
	4-*O*-Caffeoylquinic acid	OFI	Amaya-Cruz et al. (2019)
	4-*p*-Coumaroylquinic acid	OFI	Amaya-Cruz et al. (2019)
	4-Sinapoylquinic acid	OFI	Amaya-Cruz et al. (2019)
	5-*O*-Caffeoylquinic acid	OFI	Amaya-Cruz et al. (2019)
	5-*p*-Coumaroylquinic acid	OFI	Amaya-Cruz et al. (2019)
	5-Sinapoylquinic acid	OFI	Amaya-Cruz et al. (2019)
	Caffeic acid	OFI; OM	Amaya-Cruz et al. (2019), Ndhlala et al. (2007)
	Caffeic acid 4-*O*-glucoside	OFI	Amaya-Cruz et al. (2019)
	Caffeic acid hexoside	OMT	Morales et al. (2015)
	Caffeoyl tartaric acid	OFI	Amaya-Cruz et al. (2019)
	Cinnamic acid	OFI	Amaya-Cruz et al. (2019)
	Dihydrosinapic acid hexoside	OFI	Mena et al. (2018)
	Ferulic acid	OFI; OM	Amaya-Cruz et al. (2019), Ndhlala et al. (2007), Mena et al. (2018)
	Ferulic acid 4-*O*-glucoside	OFI; OMT	Amaya-Cruz et al. (2019), Morales et al. (2015)
	Isoferulic acid	OFI	Amaya-Cruz et al. (2019)
	m-Coumaric acid	OFI	Amaya-Cruz et al. (2019)
	o-Coumaric acid	OFI	Amaya-Cruz et al. (2019)
	p-Coumaric acid	OFI; OM	Amaya-Cruz et al. (2019); Ndhlala et al. (2007)
	p-Coumaric acid 4-*O*-glucoside	OFI	Amaya-Cruz et al. (2019)
	Sinapic acid	OFI	Amaya-Cruz et al. (2019)
	Sinapic acid hexoside	OFI	Mena et al. (2018)

Table 5.3 Bioactive compounds cited in *Opuntia* spp. peels.—*cont'd*

Family	Compound name	Variety	Author
Flavan-3-ols	(+)-Catechin	OMT	Guzmán-Maldonado et al. (2010)
	(−)-Epicatechin	OMT	Guzmán-Maldonado et al. (2010)
	(−)-Epicatechin 3-*O*-glucose	OFI	Amaya-Cruz et al. (2019)
Flavonols	Isorhamnetin	OFI; OL	Amaya-Cruz et al. (2019), Kuti (2004), Moussa-ayoub et al. (2011), Chougui et al. (2015)
	Isorhamnetin 3-*O*-glucoside	OFI; OSC; OMT	Amaya-Cruz et al. (2019), Moussa-ayoub et al. (2011), Melgar et al. (2017), Yeddes et al. (2013), Morales et al. (2015)
	Isorhamnetin 3-*O*-rutinoside	OFI; OMC; OSC; OMT	Moussa-Ayoub et al. (2011); Yeddes et al. (2013); Morales et al. (2015)
	Isorhamnetin 4′-*O*-glucoside	OFI	Amaya-Cruz et al. (2019)
	Isorhamentin-*O*-(deoxyhexosyl-hexoside)	OFI; OE; OMT	Melgar et al. (2017); Morales et al. (2015)
	Isorhamentin-*O*-(deoxyhexosyl-pentosyl-hexoside)	OFI	Melgar et al. (2017)
	Isorhamnetin-*O*-(di-deoxyhexosyl-hexoside)	OFI; OMT	Melgar et al. (2017); Morales et al. (2015)
	Isorhamentin-*O*-(pentosyl-hexoside)	OFI	Melgar et al. (2017)
	Kaempferide	OFI	Amaya-Cruz et al. (2019)
	Kaempferol	OFI; OL, OSP	Amaya-Cruz et al. (2019); Kuti (2004)
	Kaempferol 3-*O*-(2″-rhamnosyl-galactoside) 7-O-rhamnoside	OFI	Amaya-Cruz et al. (2019)
	Kaempferol 3-*O*-(6″-acetyl-galactoside) 7-O-rhamnoside	OFI	Amaya-Cruz et al. (2019)
	Kaempferol 3-*O*-acetyl-glucoside	OFI	Amaya-Cruz et al. (2019)
	Kaempferol 3-*O*-glucoside	OFI	Amaya-Cruz et al. (2019)

Continued

Table 5.3 Bioactive compounds cited in *Opuntia* spp. peels.—*cont'd*

Family	Compound name	Variety	Author
	Kaempferol 3-*O*-glucosyl-rhamnosyl-galactoside	OFI	Amaya-Cruz et al. (2019)
	Kaempferol 3-*O*-glucosyl-rhamnosyl-glucoside	OFI	Amaya-Cruz et al. (2019)
	Kaempferol 3-*O*-glucuronide	OFI	Amaya-Cruz et al. (2019)
	Kaempferol 3-*O*-rhamnoside	OFI	Amaya-Cruz et al. (2019)
	Kaempferol 3-*O*-rutinoside	OFI	Amaya-Cruz et al. (2019); Melgar et al. (2017); Mena et al. (2018)
	Kaempferol 3-*O*-xylosyl-glucoside	OFI	Amaya-Cruz et al. (2019)
	Kaempferol 3,7,4′-*O*-triglucoside	OFI	Amaya-Cruz et al. (2019)
	Kaempferol 7-*O*-glucoside	OFI	Amaya-Cruz et al. (2019)
	Kaempferol-*O*-(di-deoxyhexosyl-hexoside)-hexoside	OMT	Morales et al. (2015)
	Myricetin	OFI	Amaya-Cruz et al. (2019)
	Myricetin 3-*O*-glucoside	OFI	Amaya-Cruz et al. (2019)
	Myricetin 3-*O*-rhamnoside	OFI	Amaya-Cruz et al. (2019)
	Myricetin hexoside	OFI	Mena et al. (2018)
	Quercetin	OFI; OL, OSP, OSC	Amaya-Cruz et al. (2019), Kuti (2004)
	Quercetin 3-*O*-arabinoside	OFI	Amaya-Cruz et al. (2019)
	Quercetin 3-*O*-glucoside	OFI	Amaya-Cruz et al. (2019)
	Quercetin 3-*O*-glucosyl-rhamnosyl-glucoside	OFI	Amaya-Cruz et al. (2019)
	Quercetin 3-*O*-glucosyl-xyloside	OFI	Amaya-Cruz et al. (2019)
	Quercetin 3-*O*-rhamnosyl-galactoside	OFI	Amaya-Cruz et al. (2019)
	Quercetin 3-*O*-rutinoside	OFI; OSC	Melgar et al. (2017), Yeddes et al. (2013), Mena et al. (2018)

Table 5.3 Bioactive compounds cited in *Opuntia* spp. peels.—*cont'd*

Family	Compound name	Variety	Author
	Quercetin 3-*O*-xyloside	OFI	Amaya-Cruz et al. (2019)
	Quercetin 3-*O*-xylosyl-rutinoside	OFI	Amaya-Cruz et al. (2019)
	Quercetin 3,4′-*O*-diglucoside	OFI	Amaya-Cruz et al. (2019)
	Quercetin 7,4′-*O*-diglucoside	OFI	Amaya-Cruz et al. (2019)
	Quercetin-*O*-(di-deoxyhexosyl-hexoside)	OMT	Morales et al. (2015)
Flavanones	Hesperidin	OFI	Amaya-Cruz et al. (2019)
	Naringenin	OFI	Amaya-Cruz et al. (2019)
	Naringenin-hexoside	OFI	Mena et al. (2018)
	Naringin	OFI	Mena et al. (2018)
Betacyanins	15-Decarboxy-betanin	OFI	Amaya-Cruz et al. (2019)
	17-Decarboxy-betanin	OFI	Amaya-Cruz et al. (2019)
	17-Decarboxy-neobetanin	OFI	Amaya-Cruz et al. (2019)
	2-Decarboxy-betanin	OFI; OMT	Amaya-Cruz et al. (2019), Morales et al. (2015)
	2-Decarboxy-isobetanin	OMT	Morales et al. (2015)
	2-Decarboxy-neobetanin	OFI	Amaya-Cruz et al. (2019)
	2,15,17-Tridecarboxy-betanin	OFI	Amaya-Cruz et al. (2019)
	2,15,17-Tridecarboxy-neobetanin	OFI	Amaya-Cruz et al. (2019)
	Betalamic acid	OFI	Amaya-Cruz et al. (2019)
	Betanidin	OFI; OE; OSC	Amaya-Cruz et al. (2019), Melgar et al. (2017), Yeddes et al. (2013)
	Betanidin-5-*O*-β-sophoroside	OE	Melgar et al. (2017)
	Betanin	OFI: OMC; OE; OSC; OMT	Amaya-Cruz et al. (2019), Moussa-ayoub et al. (2011), Melgar et al. (2017), Yeddes et al. (2013), Morales et al. (2015)

Continued

Table 5.3 Bioactive compounds cited in *Opuntia* spp. peels.—*cont'd*

Family	Compound name	Variety	Author
	Gomphrenin I	OE	Melgar et al. (2017)
	Gomphrenin II	OFI	Amaya-Cruz et al. (2019)
	Gomphrenin III	OFI	Amaya-Cruz et al. (2019)
	Hylocerenin	OFI	Amaya-Cruz et al. (2019)
	Isobetanin	OFI; OE; OSC; OMT	Amaya-Cruz et al. (2019), Melgar et al. (2017), Yeddes et al. (2013), Morales et al. (2015)
	Isophyllocactin	OFI	Amaya-Cruz et al. (2019)
	Neobetanin	OFI	Amaya-Cruz et al. (2019)
	Phyllocactin	OFI	Amaya-Cruz et al. (2019)
Betaxanthins	3-Methoxy-tyramine-betaxanthin	OFI	Amaya-Cruz et al. (2019)
	5-Hydroxynorvaline-betaxanthin	OFI	Amaya-Cruz et al. (2019)
	Alanine-betaxanthin	OFI	Amaya-Cruz et al. (2019)
	Arginine-betaxanthin	OFI	Amaya-Cruz et al. (2019)
	Aspartic acid-betaxanthin	OFI	Amaya-Cruz et al. (2019)
	Dopa-betaxanthin	OFI	Amaya-Cruz et al. (2019)
	Dopamine-betaxanthin	OFI	Amaya-Cruz et al. (2019)
	Glutamine-betaxanthin	OFI	Amaya-Cruz et al. (2019)
	Histidine-betaxanthin	OFI	Amaya-Cruz et al. (2019)
	Indicaxanthin	OFI	Melgar et al. (2017)
	Isoleucine-betaxanthin	OFI	Amaya-Cruz et al. (2019)
	Lysine-betaxanthin	OFI	Amaya-Cruz et al. (2019)
	Methionine sulfoxide-betaxanthin	OFI	Amaya-Cruz et al. (2019)
	Methionine-betaxanthin	OFI	Amaya-Cruz et al. (2019)
	Phenylalanine-betaxanthin	OFI	Amaya-Cruz et al. (2019)
	Proline-betaxanthin	OFI	Amaya-Cruz et al. (2019)
	Proline-betaxanthin	OFI	Amaya-Cruz et al. (2019)
	Serine-betaxanthin	OFI	Amaya-Cruz et al. (2019)
	Threonine-betaxanthin	OFI	Amaya-Cruz et al. (2019)
	Tryptophan-betaxanthin	OFI	Amaya-Cruz et al. (2019)

Table 5.3 Bioactive compounds cited in *Opuntia* spp. peels.—*cont'd*

Family	Compound name	Variety	Author
	Tyramine-betaxanthin	OFI	Amaya-Cruz et al. (2019)
	Tyrosine-betaxanthin	OFI	Amaya-Cruz et al. (2019)
	Valine-betaxanthin	OFI	Amaya-Cruz et al. (2019)
	γ-Aminobutyric acid-betaxanthin	OFI	Amaya-Cruz et al. (2019)
Carotenoids	Antheraxanthin	OFI	Amaya-Cruz et al. (2019), Cano et al. (2017)
	Astaxanthin	OFI	Amaya-Cruz et al. (2019)
	Bixin	OFI	Amaya-Cruz et al. (2019)
	Canthaxanthin	OFI	Amaya-Cruz et al. (2019)
	Flavoxanthin	OFI	Amaya-Cruz et al. (2019)
	Lutein	OFI	Cano et al. (2017)
	Lycopene	OFI	Cano et al. (2017)
	Neoxanthin	OFI	Amaya-Cruz et al. (2019), Cano et al. (2017)
	Neurosporene	OFI	Amaya-Cruz et al. (2019)
	Norbixin	OFI	Amaya-Cruz et al. (2019)
	Physalien	OFI	Amaya-Cruz et al. (2019)
	Phytofluene	OFI	Amaya-Cruz et al. (2019)
	Violaxanthin	OFI	Amaya-Cruz et al. (2019), Cano et al. (2017)
	Zeaxanthin	OFI	Cano et al. (2017)
	α-Carotene	OFI	Amaya-Cruz et al. (2019)
	α-Cryptoxanthin	OFI	Amaya-Cruz et al. (2019), Cano et al. (2017)
	β-Apo-8′-Carotenal	OFI	Amaya-Cruz et al. (2019)
	β-Carotene	OFI	Amaya-Cruz et al. (2019), Cano et al. (2017)
	β-Cryptoxanthin	OFI	Cano et al. (2017)
	ζ-Carotene	OFI	Amaya-Cruz et al. (2019)
Phytosterols	Campesteryl β-d-glucoside	OFI	Amaya-Cruz et al. (2019)
	Fucosterol	OFI	Amaya-Cruz et al. (2019)
	Lanosterol	OFI	Amaya-Cruz et al. (2019)
	Sitosteryl β-d-glucoside	OFI	Amaya-Cruz et al. (2019)
	Stigmasteryl β-d-glucoside	OFI	Amaya-Cruz et al. (2019)
Chlorophylls	Chlorophyll *a*	OFI	Cano et al. (2017)
	Chlorophyll *b*	OFI	Cano et al. (2017)

Continued

Table 5.3 Bioactive compounds cited in *Opuntia* spp. peels.—*cont'd*

Family	Compound name	Variety	Author
Lignin	Guaiacyl(8-*O*-4) syringyl(8–8) guaiacyl-hexoside	OFI	Mena et al. (2018)
	Guaiacyl(t8-*O*-4) guaiacyl-hexoside	OFI	Mena et al. (2018)
	Secoisolariciresinol-hexoside	OFI	Mena et al. (2018)
	Syringaresinol	OFI	Mena et al. (2018)
	Syringyl(t8-*O*-4) guaiacyl	OFI	Mena et al. (2018)

OE, Opuntia engelmannii; *OFI*, Opuntia ficus-indica; *OL*, Opuntia lindheimeri; *OM*, Opuntia megacantha; *OMC*, Opuntia macrorhiza; *OMT*, Opuntia matudae; *OSC*, Opuntia stricta; *OSP*, Opuntia streptacantha.

3.4.2 Industrial processing: integrated processes for the recovery of bioactive compounds

Regarding phenolic composition, Aguirre et al. (2013) investigated the extraction of polyphenols from *Opuntia ficus-indica* skin by using a reflux system. The optimum conditions for the extraction of polyphenols were 45% of ethanol, 80°C, and 2 h of extraction. Higher temperatures led to an increase of the extraction of polyphenols but a decrease of the antioxidant potential.

The potential of an integrated membrane process for producing cactus pear juice concentrate, with high nutritional value, has been investigated by Cassano et al. (2007). The process included an initial clarification of the depectinized juice by UF. The clarified juice was then concentrated by OD from 11 °Brix of total soluble solids (TSS) up to 61 °Brix. The content of Vitamin C, glutamic acid, citric acid, and betalains was very well preserved during the process independently by the TSS content. The evaluation of the TAA in the concentrated samples confirmed the validity of the process in preserving bioactive compounds of the fresh juice.

Therefore, the OD retentate was considered suitable for nutraceutical applications (i.e., nutritional supplements) or as coloring foodstuff due to its high concentration of betalains.

Cassano et al. (2010) evaluated also the effect of polyvinylidenfluoride flat-sheet MF and UF membranes (0.20 μm as pore size and 200 kDa as MWCO, respectively) on the physicochemical composition of the cactus pear juice. The clarified juice showed physicochemical and nutritional properties similar to those of the fresh juice except for the absence of suspended solids and betalains, which were retained by the selected membranes. Accordingly, the retentate fraction can be used as a raw material to extract betalains or directly as functional food.

More recently, an integrated membrane-based process aimed to purify, fractionate, and concentrate betacyanins in cactus pear juice has been investigated by

Table 5.4 Extraction systems and final detection analysis using *Opuntia* spp. by-products.

Technology of extraction	Final detection technique	Solvent	Author
Homogenization	HPLC-DAD	Methanol:water (50/50)	Kuti (2004)
	HPLC-DAD-MS	Tetrahydrofuran	Cano et al. (2017)
Maceration	UPLC-QTOF-MS[e]	Methanol:water (50/50) Acetone:water (70/30)	Amaya-Cruz et al. (2019)
	HPLC-UV-Vis	Methanol:water (50/50)	Ndhlala et al. (2007)
	HPLC-DAD	Methanol:water (70/30)	Guzmán-Maldonado et al. (2010)
	UPLC-DAD-ESI/MS^n	Ethanol:water (80/20)	Melgar et al. (2017)
	HPLC-QTRAP-MS	Methanol:water (80/20)	Morales et al. (2015)
	Spectrophotometry	Methanol:water (66/34)	Diaz-Vela et al. (2013)
	UPLC-DAD-ESI/MS^n	Ethanol:water (70/30)	Chougui et al. (2015)
Microwave-assisted extraction	Scanning electron microscope	Water	Han et al. (2016)
Pulsed electric field	HPLC-DAD	Water	Koubaa et al. (2016)
Reflux extraction	Spectrophotometry	Ethanol:water	Aguirre-Joya et al. (2013)
Supercritical fluids extraction	UPLC-QTOF-MS	Carbon dioxide	Koubaa et al. (2016)
Ultrasound-assisted extraction	HPLC-DAD	Methanol:water (80/20)	Moussa-Ayoub et al. (2011)
	HPLC-DAD	Methanol:water (70/30)	Moussa-ayoub et al. (2011)
	RP-HPLC-ESI-MS	Methanol:acetic acid (99/1)	Yeddes et al. (2013)
	UHPLC-ESI-MS	Methanol:water (80/20)	Mena et al. (2018)

HPLC, *High-performance liquid chromatography;* UHPLC, *Ultrahigh-performance liquid chromatography;* RP, *Reverse phase;* UPLC, *Ultra-performance liquid chromatography;* DAD, *Diode array detector;* ESI, *Electrospray ionization;* MS, *Mass spectrometry;* UV-Vis, *Ultraviolet visible spectra;* QTOF, *Quadrupole time of flight;* QTRAP, *Triple quadrupole linear ion trap.*

Tamba et al. (2019). The proposed design is based on the MF of depectinized juice to promote the removal of insoluble solids, followed by UF or NF treatment for solute separation. By selecting the appropriate UF or NF membrane and operating pressure, the retention of solutes can be modulated to promote the concentration of all the solutes or, alternatively, the purification of betacyanins with respect to the total dry matter.

4. Conclusions

Next generations will have to focus on different issues: (1) population increment that leads to a higher food needs; (2) healthy authorities and consumers require each time higher food standards in terms of nutritional values and safety of food products; and (3) solve environmental problems related to waste generation along the food chain.

This chapter emphasizes in the need of the use of green and clean technologies for the waste treatment and for the recovery and obtaining of biocompounds from food wastes. Besides, some case studies have been described as successful source of biocomponents that can be recovered from various vegetable sources.

References

Abbas, Z.K., Saggu, S., Sakeran, M.I., Zidan, N., Rehman, H., Ansari, A.A., 2014. Phytochemical, antioxidant and mineral composition of hydroalcoholic extract of chicory (*Cichorium intybus* L.) leaves. Saudi Journal of Biological Sciences 22 (3), 322–326.

Agcam, E., Akyıldız, A., Balasubramaniam, V.M., 2017. Optimization of anthocyanins extraction from black carrot pomace with thermosonication. Food Chemistry 237, 461–470.

Aguirre, J.J., De La Garza, T.H., Zugasti, C.A., Belmares, C.R., Aguilar, C.N., 2013. The optimization of phenolic compounds extraction from cactus pear (*Opuntia ficus-indica*) skin in a reflux system using response surface methodology. Asian Pacific Journal of Tropical Biomedicine 3 (6), 436–442.

Alvarez, M.V., Cabred, S., Ramirez, C.L., Fanovich, M.A., 2019. Valorization of an agroindustrial soybean residue by supercritical fluid extraction of phytochemical compounds. The Journal of Supercritical Fluids 143, 90–96.

Amaya-Cruz, D.M., Pérez-Ramírez, I.F., Delgado-García, J., Mondragón-Jacobo, C., Dector-Espinoza, A., Reynoso-Camacho, R., 2019. An integral profile of bioactive compounds and functional properties of prickly pear (*Opuntia ficus indica* L.) peel with different tonalities. Food Chemistry 278, 568–578.

Araújo, M., Pimentel, F.B., Alves, R.C., Oliveira, M.B.P.P., 2015. Phenolic compounds from olive mill wastes: health effects, analytical approach and application as food antioxidants. Trends in Food Science & Technology 45 (2), 200–211.

Bagherian, H., Zokaee Ashtiani, F., Fouladitajar, A., Mohtashamy, M., 2011. Comparisons between conventional, microwave- and ultrasound-assisted methods for extraction of pectin from grapefruit. Chemical Engineering and Processing: Process Intensification 50 (11), 1237–1243.

Balakrishnan, S., Natarajan, B., Arumugam, B., Hariram, B.V., 2011. Pharmacognostical evaluation of *Amaranthus spinosus* L. Pharmacognosy Journal 3 (19), 13–18.

Barba, F.J., Putnik, P., Bursać Kovačević, D., Poojary, M.M., Roohinejad, S., Lorenzo, J.M., Koubaa, M., 2017. Impact of conventional and non-conventional processing on prickly pear (*Opuntia* spp.) and their derived products: from preservation of beverages to valorization of by-products. Trends in Food Science & Technology 67, 260–270.

Bazzarelli, F., Piacentini, E., Poerio, T., Mazzei, R., Cassano, A., Giorno, L., 2016. Advances in membrane operations for water purification and biophenols recovery/valorization from OMWWs. Journal of Membrane Science 497, 402–409.

Boukroufa, M., Boutekedjiret, C., Petigny, L., Rakotomanomana, N., Chemat, F., 2015. Biorefinery of orange peels waste: a new concept based on integrated green and solvent free extraction processes using ultrasound and microwave techniques to obtain essential oil, polyphenols and pectin. Ultrasonics Sonochemistry 24, 72–79.

Bot, F., Verkerk, R., Mastwijk, H., Anese, M., Fogliano, V., Capuano, E., 2018. The effect of pulsed electric fields on carotenoids bioaccessibility: the role of tomato matrix. Food Chemistry 240, 415–421.

Bvenura, C., Sivakumar, D., 2017. The role of wild fruits and vegetables in delivering a balanced and healthy diet. Food Research International 99 (1), 15–30.

Caleja, C., Santos-Buelga, C., Ciric, A., Antonio, A.L., Oliveira, M.B.P.P., Barros, L., Soković, M., 2014. Foeniculum vulgare Mill. as natural conservation enhancer and health promoter by incorporation in cottage cheese. Journal of Functional Foods 12, 428–438.

Cano, M.P., Gómez-Maqueo, A., García-Cayuela, T., Welti-Chanes, J., 2017. Characterization of carotenoid profile of Spanish Sanguinos and Verdal prickly pear (*Opuntia ficus-indica*, spp.) tissues. Food Chemistry 237, 612–622.

Carciochi, R.A., D'Alessandro, L.G., Vauchel, P., Rodriguez, M.M., Nolasco, S.M., Dimitrov, K., 2017. Valorization of agrifood by-products by extracting valuable bioactive compounds using green processes. Ingredients Extraction by Physicochemical Methods in Food 191–228.

Carvalho, A.M., Sánchez-Mata, M.C., Molina, M., Morales, P., Cámara, M., Ferreira, I.C.F.R., 2011. Tocopherol composition and antioxidant activity of Spanish wild vegetables. Genetic Resources and Crop Evolution 59 (5), 851–863.

Cassano, A., Conidi, C., Timpone, R., D'Avella, M., Drioli, E., 2007. A membrane-based process for the clarification and the concentration of the cactus pear juice. Journal of Food Engineering 80 (3), 914–921.

Cassano, A., Conidi, C., Drioli, E., 2010. Physico-chemical parameters of cactus pear (*Opuntia ficus-indica*) juice clarified by microfiltration and ultrafiltration processes. Desalination 250 (3), 1101–1104.

Cassano, A., Conidi, C., Drioli, E., 2011. Comparison of the performance of UF membranes in olive mill wastewaters treatment. Water Research 45 (10), 3197–3204.

Cassano, A., Conidi, C., Giorno, L., Drioli, E., 2013. Fractionation of olive mill wastewaters by membrane separation techniques. Journal of Hazardous Materials 248–249, 185–193.

Cassano, A., Conidi, C., Ruby-Figueroa, R., 2014. Recovery of flavonoids from orange press liquor by an integrated membrane process. Membranes 4 (3), 509–524.

Chougui, N., Djerroud, N., Naraoui, F., Hadjal, S., Aliane, K., Zeroual, B., Larbat, R., 2015. Physicochemical properties and storage stability of margarine containing *Opuntia ficus-indica* peel extract as antioxidant. Food Chemistry 173, 382–390.

Conidi, C., Cassano, A., Drioli, E., 2011. A membrane-based study for the recovery of polyphenols from bergamot juice. Journal of Membrane Science 375 (1–2), 182–190.

Conidi, C., Cassano, A., Drioli, E., 2012. Recovery of phenolic compounds from orange press liquor by nanofiltration. Food and Bioproducts Processing 90 (C4), 867–874.

Conidi, C., Mazzei, R., Cassano, A., Giorno, L., 2014. Integrated membrane system for the production of phytotherapics from olive mill wastewaters. Journal of Membrane Science 454, 322–329.

Conidi, C., Rodriguez-Lopez, A.D., Garcia-Castello, E.M., Cassano, A., 2015. Purification of artichoke polyphenols by using membrane filtration and polymeric resins. Separation and Purification Technology 144, 153–161.

Conidi, C., Drioli, E., Cassano, A., 2018. Membrane-based agro-food production processes for polyphenol separation, purification and concentration. Current Opinion in Food Science 23, 149–164.

Cvjetko Bubalo, M., Vidovi, S., Radojcic Redovnikovic, I., Jokic, S., 2018. New perspective in extraction of plant biologically active compounds by green solvents. Food and Bioproducts Processing Journal 109, 52–73.

Delgado-Vargas, F., Paredes-Lopez, O., 2003. Natural colorants for food and nutraceutical uses. Trends in Food Science & Technology 14. https://doi.org/10.1016/S0924-2244(03)00076-1.

Diaz-Vela, J., Totosaus, A., Cruz-Guerrero, A.E., Pérez-Chabela, M.L., 2013. In vitro evaluation of the fermentation of added-value agroindustrial by-products: cactus pear (*Opuntia ficus-indica* L.) peel and pineapple (*Ananas comosus*) peel as functional ingredients. International Journal of Food Science and Technology 48 (7). https://doi.org/10.1111/ijfs.12113.

Directive 2008/98/EC of The European Parliament and of the council on waste and repealing certain directives. Official Journal of the European Union, 312, 3-30. https://doi.org/10.1016/j.jenvman.2018.12.041.

Eh, A.L.S., Teoh, S.G., 2012. Novel modified ultrasonication technique for the extraction of lycopene from tomatoes. Ultrasonics Sonochemistry 19 (1), 151–159.

El-Kader, M.A., El-Mawla, A., Mohamed, H.M., Ibraheim, Z., 2006. Phytochemical and biological studies of *Emex spinosa* (L.) Campd. growing in Egypt. Bulletin of Pharmaceutical Sciences 29.

Enrique de Briano, A.E., Acciaresi, H.A., Briano, J.A., 2013. Establishment, dispersal, and prevalence of *Rhinocyllus conicus* (Coleoptera: Curculionidae), a biological control agent of thistles, *Carduus* species (*Asteraceae*), in Argentina, with experimental information on its damage. Biological Control 67 (2), 186–193.

Farkhondeh, T., Samarghandian, S., 2019. The therapeutic effects of *Portulaca oleracea* L. in hepatogastric disorders. Gastroenterología y Hepatología 42 (2), 127–132.

Faustino, M.V., Pinto, D.C.G.A., Gonçalves, M.J., Salgueiro, L., Silveira, P., Silva, A.M.S., 2018. Calendula L. species polyphenolic profile and in vitro antifungal activity. Journal of Functional Foods 45, 254–267.

Fava, F., Totaro, G., Diels, L., Reis, M., Duarte, J., Carioca, B.O., Poggi-Varaldo, H.M., Ferreira, S.B., 2015. Biowaste biorefinery in Europe: opportunities and research & development needs. New Biotechnology 32, 100–108.

Ferrentino, G., Morozova, K., Mosibo, O.K., Ramezani, M., Scampicchio, M., 2018. Biorecovery of antioxidants from apple pomace by supercritical fluid extraction. Journal of Cleaner Production 186, 253–261.

Fidalgo, A., Ciriminna, R., Carnaroglio, D., Tamburino, A., Cravotto, G., Grillo, G., Ilharco, L.M., Pagliaro, M., 2016. Eco-Friendly extraction of pectin and essential oils from orange and lemon peels. ACS Sustainable Chemical Engineering 4 (4), 2243–2251.

Food And Agricultural Organization (FAO), 2016. Citrus Fruits Statistics 2017. Data from 2008-2016. http://www.fao.org/economic/est/est-commodities/citrus-fruit/it/. Accessed on 4th February 2019.

Food and Agricultural Organization (FAO), 2009. FAO's Director-general on how to feed the world in 2050. Population and Development Review 35 (4), 837−839.

Food and Agricultural Organization (FAO), 2011. Global Food Losses and Food Waste: Extent, Causes and Prevention. Food and Agriculture Organization of the United Nations, Rome.

Galanakis, C.M., 2012. Recovery of high added-value components from food wastes: conventional, emerging technologies and commercialized applications. Trends in Food Science & Technology 26 (2), 68−87.

Galanakis, C.M., Castro-Muñoz, R., Cassano, A., Conidi, C., 2016. Recovery of high-added-value compounds from food waste by membrane technology. In: Membrane Technologies for Biorefining. Woodhead Publishing. https://doi.org/10.1016/B978-0-08-100451-7.00008-6.

Galanakis, C.M., 2018. Food waste recovery: prospects and opportunities. In: Sustainable Food Systems from Agriculture to Industry. Elsevier Inc. https://doi.org/10.1016/B978-0-12-811935-8.00012-3.

Gallucci, M.N., Aiassa Martínez, I.M., Páez, P.L., Coronado, E.A., Ferreyra Maillard, A.P.V., Dalmasso, P.R., Pannunzio Miner, E.V., 2017. Silver nanoparticles from leafy green extract of Belgian endive (*Cichorium intybus* L. var. sativus): biosynthesis, characterization, and antibacterial activity. Materials Letters 197, 98−101.

Garcia-Castello, E., Cassano, A., Criscuoli, A., Conidi, C., Drioli, E., 2010. Recovery and concentration of polyphenols from olive mill wastewaters by integrated membrane system. Water Research 44 (13), 3883−3892.

Garcia-Castello, E.M., Rodriguez-Lopez, A.D., Mayor, L., Ballesteros, R., Conidi, C., Cassano, A., 2015. Optimization of conventional and ultrasound assisted extraction of flavonoids from grapefruit (*Citrus paradise* L.) solid wastes. Lebensmittel-Wissenschaft und -Technologie- Food Science and Technology 64 (2), 1114−1122.

García-Cruz, L., Dueñas, M., Santos-Buelgas, C., Valle-Guadarrama, S., Salinas-Moreno, Y., 2017. Betalains and phenolic compounds profiling and antioxidant capacity of pitaya (*Stenocereus* spp.) fruit from two species (*S. pruinosus* and *S. stellatus*). Food Chemistry 234, 111−118.

Garrido, T., Gizdavic-Nikolaidis, M., Leceta, I., Urdanpilleta, M., Guerrero, P., de la Caba, K., Kilmartin, P.A., 2019. Optimizing the extraction process of natural antioxidants from chardonnay grape marc using microwave-assisted extraction. Waste Management 88, 110−117.

Georgiev, Y., Ognyanov, M., Yanakieva, I., Kussovski, V., Kratchanova, M., 2012. Isolation, characterization and modification of citrus pectins. Journal of Bioscience and Biotechnology 1, 223−233.

Ghosh, S., Gillis, A., Sheviryov, J., Levkov, K., Golberg, A., 2019. Towards waste meat biorefinery: extraction of proteins from waste chicken meat with non-thermal pulsed electric fields and mechanical pressing. Journal of Cleaner Production 208, 220−231.

Görgüç, A., Bircan, C., Yılmaz, F.M., 2019. Sesame bran as an unexploited by-product: effect of enzyme and ultrasound-assisted extraction on the recovery of protein and antioxidant compounds. Food Chemistry 283, 637−645.

Guijarro-Real, C., Rodríguez-Burruezo, A., Prohens, J., Raigón, M.D., Fita, A., 2019. HS-SPME analysis of the volatiles profile of water celery (*Apium nodiflorum*), a wild vegetable with increasing culinary interest. Food Research International 121, 765−775.

Guzmán-Maldonado, S.H., Herrera-Hernández, G., Hernández-López, D., Reynoso-Camacho, R., Guzmán-Tovar, A., Vaillant, F., Brat, P., 2010. Physicochemical, nutritional and functional characteristics of two underutilised fruit cactus species (Myrtillocactus) produced in Central Mexico. Food Chemistry 121 (2), 381–386.

Han, Y.-L., Gao, J., Yin, Y.-Y., Jin, Z.-Y., Xu, X.-M., Chen, H.-Q., 2016. Extraction optimization by response surface methodology of mucilage polysaccharide from the peel of *Opuntia dillenii* haw. fruits and their physicochemical properties. Carbohydrate Polymers 151, 381–391.

Hayat, K., Hussain, S., Abbas, S., Farooq, U., Ding, B., Xia, S., Jia, C., Zhang, X., Xia, W., 2009. Optimized microwave-assisted extraction of phenolic acids from citrus Mandarin peels and evaluation of antioxidant activity in vitro. Separation and Purification Technology 70 (1), 63–70.

Herbach, K.M., Stintzing, F.C., Carle, R., 2006. Betalain stability and degradation – structural and chromatic aspects. Journal of Food Science 71 (4), 41–50.

Herrera, M.C., Luque de Castro, M.D., 2005. Ultrasound-assisted extraction of phenolic compounds from strawberries prior to liquid chromatographic separation and photodiode array ultraviolet detection. Journal of Chromatography A 1100 (1), 1–7.

Hilou, A., Nacoulma, O.G., Guiguemde, T.R., 2006. In vivo antimalarial activities of extracts from *Amaranthus spinosus* L. and *Boerhaavia erecta* L. in mice. Journal of Ethnopharmacology 103 (2), 236–240.

Iamonico, D., 2019. Remarks on the subfam. Betoideae (Chenopodiaceae/Amaranthaceae sensu APGIV). Phytotaxa 400. https://doi.org/10.11646/phytotaxa.400.3.3.

ISEKI Food Association, 2012. Special Interest Group, SIG-5: Food Waste Recovery. https://www.iseki-food.net/general/sigs/sig5.

Kaderides, K., Papaoikonomou, L., Serafim, M., Goula, A.M., 2019. Microwave-assisted extraction of phenolics from pomegranate peels: optimization, kinetics, and comparison with ultrasounds extraction. Chemical Engineering and Processing – Process Intensification 137, 1–11.

Kalogerakis, N., Politi, M., Foteinis, S., Chatzisymeon, E., Mantzavinos, D., 2013. Recovery of antioxidants from olive mill wastewaters: a viable solution that promotes their overall sustainable management. Journal of Environmental Management 128, 749–758.

Khan, M.K., Abert-Vian, M., Fabiano-Tixier, A.S., Dangles, O., Chemat, F., 2010. Ultrasound-assisted extraction of polyphenols (flavanone glycosides) from orange (*Citrus sinensis* L.) peel. Food Chemistry 119 (2), 851–858.

Klen, T.J., Vodopivec, B.M., 2011. Ultrasonic extraction of phenols from olive mill wastewater: comparison with conventional methods. Journal of Agricultural and Food Chemistry 59 (24), 12725–12731.

Koubaa, M., Barba, F.J., Grimi, N., Mhemdi, H., Koubaa, W., Boussetta, N., Vorobiev, E., 2016. Recovery of colorants from red prickly pear peels and pulps enhanced by pulsed electric field and ultrasound, 37, 336–344.

Kuti, J.O., 2004. Antioxidant compounds from four *Opuntia* cactus pear fruit varieties. Food Chemistry 85 (4), 527–533.

Lavilla, I., Bendicho, C., 2017. Fundamentals of ultrasound-assisted extraction. In: Water Extraction of Bioactive Compounds: From Plants to Drug Development. Elsevier Inc. https://doi.org/10.1016/B978-0-12-809380-1.00011-5.

Levorato, S., Dominici, L., Fatigoni, C., Zadra, C., Pagiotti, R., Moretti, M., Villarini, M., 2018. In vitro toxicity evaluation of estragole-containing preparations derived from

Foeniculum vulgare Mill. (fennel) on HepG2 cells. Food and Chemical Toxicology 111, 616–622.

Lin, J.-H., Lin, M.-F., 1997. Tannins and related compounds from *Erodium moschatum* (L.) L'Her. Journal of Food and Drug Analysis 5.

Manna, L., Bugnone, C.A., Banchero, M., 2015. Valorization of hazelnut, coffee and grape wastes through supercritical fluid extraction of triglycerides and polyphenols. The Journal of Supercritical Fluids 104, 204–211.

Maran, J.P., Sivakumar, V., Thirugnanasambandham, K., Sridhar, R., 2013. Optimization of microwave assisted extraction of pectin from orange peel. Carbohydrate Polymers 97 (2), 703–709.

Maran, J.P., Swathi, K., Jeevitha, P., Jayalakshmi, J., Ashvini, G., 2015. Microwave-assisted extraction of pectic polysaccharide from waste mango peel. Carbohydrate Polymers 123, 67–71.

Martín, Á., Navarrete, A., 2018. Microwave-assisted process intensification techniques. Current Opinion in Green and Sustainable Chemistry 11, 70–75.

Martins, N., Ferreira, I.C.F.R., 2017. Mountain food products: a broad spectrum of market potential to be exploited. Trends in Food Science & Technology 67, 12–18.

Maxia, A., Falconieri, D., Piras, A., Porcedda, S., Marongiu, B., Frau, M.A., Salgueiro, L., 2012. Chemical composition and antifungal activity of essential oils and supercritical CO_2 extracts of *Apium nodiflorum* (L.) Lag. Mycopathologia 174 (1), 61–67.

Melgar, B., Dias, M.I., Ciric, A., Sokovic, M., Garcia-Castello, E.M., Rodriguez-Lopez, A.D., Ferreira, I., 2017. By-product Recovery of Opuntia Spp. Peels: Betalainic and Phenolic Profiles and Bioactive Properties. https://doi.org/10.1016/j.indcrop.2017.06.011.

Mena, P., Tassotti, M., Andreu, L., Nuncio-Jáuregui, N., Legua, P., Del Rio, D., Hernández, F., 2018. Phytochemical characterization of different prickly pear (*Opuntia ficus-indica* (L.) Mill.) cultivars and botanical parts: UHPLC-ESI-MSnmetabolomics profiles and their chemometric analysis. Food Research International 108, 301–308.

Morales, P., Barros, L., Ramírez-Moreno, E., Santos-Buelga, C., Ferreira, I.C.F.R., 2014. Exploring xoconostle by-products as sources of bioactive compounds. Food Research International 65, 437–444.

Morales, P., Barros, L., Ramírez-Moreno, E., Santos-Buelga, C., Ferreira, I.C.F.R., 2015. Xoconostle fruit (*Opuntia matudae* Scheinvar cv. Rosa) by-products as potential functional ingredients. Food Chemistry 185, 289–297.

Morales-Martínez, Y., López-Cuellar, M. del R., Chavarría-Hernández, N., Rodríguez-Hernández, A.I., 2018. Rheological behaviour of acetylated pectins from cactus pear fruits (*Opuntia albicarpa* and *O. matudae*). Food Hydrocolloids 85, 110–119.

Moussa-Ayoub, T.E., El-Samahy, S.K., Kroh, L.W., Rohn, S., 2011. Identification and quantification of flavonol aglycons in cactus pear (*Opuntia ficus indica*) fruit using a commercial pectinase and cellulase preparation. Food Chemistry 124 (3), 1177–1184.

Nayak, A., Bhushan, B., 2019. An overview of the recent trends on the waste valorization techniques for food wastes. Journal of Environmental Management 233, 352–370.

Ndhlala, A.R., Kasiyamhuru, A., Mupure, C., Chitindingu, K., Benhura, M.A., Muchuweti, M., 2007. Phenolic composition of Flacourtia indica, *Opuntia megacantha* and *Sclerocarya birrea*. Food Chemistry 103 (1), 82–87.

Obied, H.K., Allen, M.S., Bedgood, D.R., Prenzler, P.D., Robards, K., Stockmann, R., 2005. Bioactivity and analysis of biophenols recovered from olive mill waste. Journal of Agricultural and Food Chemistry 53 (4), 823–827.

Paniwnyk, L., Alarcon-Rojo, A., Rodriguez-Figueroa, J.C., Toma, M., 2017. The use of ultrasound as an enhancement aid to food extraction. In: Ingredients Extraction by Physicochemical Methods in Food. Academic Press. https://doi.org/10.1016/B978-0-12-811521-3.00010-7.

Paraskeva, C.A., Papadakis, V.G., Tsarouchi, E., Kanellopoulou, D.G., Koutsoukos, P.G., 2007. Membrane processing for olive mill wastewater fractionation. Desalination 213 (1–3), 218–229.

Peter, K., Gandhi, P., 2017. Rediscovering the therapeutic potential of *Amaranthus* species : a review. Egyptian Journal of Basic and Applied Sciences 4 (3), 196–205.

Płotka-Wasylka, J., Rutkowska, M., Owczarek, K., Tobiszewski, M., Namieśnik, J., 2017. Extraction with environmentally friendly solvents. TRAC Trends in Analytical Chemistry 91, 12–25.

Quintin, D., Garcia-Gomez, P., Ayuso, M., Sanmartin, A.M., 2019. Active biocompounds to improve food nutritional value. Trends in Food Science & Technology 84, 19–21.

Qurishi, M.A., Dar, B.A., Bhat, B.A., Rather, M.A., Sofi, S.N., 2012. *Foeniculum vulgare*: a comprehensive review of its traditional use, phytochemistry, pharmacology, and safety. Arabian Journal of Chemistry 9, S1574–S1583.

Radrigán, R., Aqueveque, P., Ocampo, M., 2017. Extraction and use of functional plant ingredients for the development of functional foods. In: Ingredients Extraction by Physico-Chemical Methods in Food. Elsevier Inc. https://doi.org/10.1016/B978-0-12-811521-3/00007-7.

Raghavarao, K.S.M.S., Madhusudhan, M.C., Hrishikesh Tavanandi, A., Niranjan, K., 2014. Athermal membrane processes for the concentration of liquid foods and natural colors. Emerging Technologies for Food Processing 12. https://doi.org/10.1016/B978-0-12-411479-1.00012-7.

Renard, C.M.G.C., 2018. Extraction of bioactives from fruit and vegetables: state of the art and perspectives. Lebensmittel-Wissenschaft und -Technologie- Food Science and Technology 93, 390–395.

Roberfroid, M., 2011. Defining functional foods and associated claims. In: Functional Foods, second ed. https://doi.org/10.1533/9780857092557.1.3

Saffarzadeh-Matin, S., Khosrowshahi, F.M., 2017. Phenolic compounds extraction from Iranian pomegranate (*Punica granatum*) industrial waste applicable to pilot plant scale. Industrial Crops and Products 108, 583–597.

Sawicki, T., Bączek, N., Wiczkowski, W., 2016. Betalain profile, content and antioxidant capacity of red beetroot dependent on the genotype and root part. Journal of Functional Foods 27, 249–261.

Stintzing, F.C., Herbach, K.M., Mosshammer, M.R., Carle, R., Yi, W., Sellappan, S., Felker, P., 2005. Color, betalain pattern, and antioxidant properties of cactus pear (*Opuntia* spp.) clones. Journal of Agricultural and Food Chemistry 53 (2), 442–451.

Tamba, A., Servent, A., Mertz, C., Cissé, M., Dornier, M., 2019. Coupling of pressure-driven membrane technologies for concentrating, purifying and fractionizing betacyanins in cactus pear (*Opuntia dillenii* Haw.) juice. Innovative Food Science & Emerging Technologies 52, 244–255.

Tardugno, R., Pozzebon, M., Beggio, M., Del Turco, P., Pojana, G., 2018. Polyphenolic profile of Cichorium intybus L. endemic varieties from the Veneto region of Italy. Food Chemistry 266, 175–182.

United Nations, 2015. Transforming Our World: The 2030 Agenda for Sustainable Development. NY, USA. https://sustainabledevelopment.un.org/post2015/transformingourworld.

USEPA, United States Envirormental Protection Agency, 2001. Waste Avoidance and Resource Recovery Act 2001. https://www.epa.nsw.gov.au/your-environment/recycling-and-reuse/warr-strategy/the-waste-hierarchy.

Valdés, A., León, C., 2017. Foodomics evaluation of bioactive compounds in foods. TRAC Trends in Analytical Chemistry 96, 2–13.

Vilkhu, K., Mawon, R., Simons, L., Bates, D., 2008. Applications and opportunities for ultrasound assisted extraction in the food industry - a review. Science and Emerging Technologies 9 (2), 161–169.

Wang, J., Vanga, S.K., Raghavan, V., 2019. High-intensity ultrasound processing of kiwifruit juice: effects on the ascorbic acid, total phenolics, flavonoids and antioxidant capacity. Lebensmittel-Wissenschaft und -Technologie- Food Science and Technology 107, 299–307.

WOS, Web of Science, 2019. Opuntia Fruit Database on Web of Science. Retrieved from: https://apps.webofknowledge.com/UA_AdvancedSearch_input.do?product=UA&search_mode=AdvancedSearch&replaceSetId=&goToPageLoc=SearchHistoryTableBanner&SID=E1XpfRWD9Opu5Tj5EcL&errorQid=1#SearchHistoryTableBanner.

Yeddes, N., Chérif, J.K., Guyot, n S., Sotin, H., Ayadi, M.T., 2013. Comparative study of antioxidant power, polyphenols, flavonoids and betacyanins of the peel and pulp of three tunisian *Opuntia* forms. Antioxidants 2, 37–51.

CHAPTER

6 Use of by-products in edible coatings and biodegradable packaging materials for food preservation

A. Chiralt[1], C. Menzel[2], E. Hernandez-García[1], S. Collazo[1], C. Gonzalez-Martinez, PhD [1]

[1]*Instituto de Ingeniería de Alimentos para el desarrollo, Universitat Politècnica de València, Valencia, Spain;* [2]*Division of Glycoscience, Department of Chemistry, School of Engineering Sciences in Chemistry, Biotechnology and Health, KTH Royal Institute of Technology, AlbaNova University Centre, Stockholm, Sweden*

A large part of the by-products of plant and animal origin are sources of different valuable compounds, such as polysaccharides or proteins, which are useful materials for developing edible or biodegradable coatings and film formulations. Nowadays, this has acquired great importance in the development of new materials for substituting, at least partially, the traditional food packaging plastics, which show significant shortcomings in terms of their environmental impact. This also implies the use of renewable resources instead of the limited petroleum reserves.

Alternative and novel materials, which possess commercial potential, are those derived from underused food products, renewable resources, and the valorization of agro-industrial and marine wastes. The exploitation of these materials additionally provides added value to the waste products and represents an attractive alternative to the nonrenewable sources, in the framework of circular economy. In addition, most of them are not just biodegradable, but are also edible, or can even be considered as active compounds (e.g., with antimicrobial or antioxidant properties), thereby contributing to the development of new sustainable packaging strategies for food preservation. In this chapter, these alternative materials are explored, and the benefits and drawbacks are addressed.

1. Importance of using by-products for the development of packaging materials

There is increasing EU regulatory pressure, together with indirect demands via consumer groups on the food industry and packaging manufacturers, for the

Sustainability of the Food System. https://doi.org/10.1016/B978-0-12-818293-2.00006-9

development of "environmentally friendly" packaging systems. Nevertheless, this current trend in new food packaging development should not only be natural and "environmentally friendly," but also functional and cost-effective. As food manufacturers require packaging materials to be food grade, the enhancement of the product shelf-life stability and safety and the reduction or replacement with alternative biodegradable forms would clearly allow an improvement in the overall operating costs while reducing the waste streams. Thus, research development in the area of edible coatings and biodegradable films is, nowadays, a key and unique field of exploration within food packaging, which possesses an enormous commercial potential.

In addition, both from an economic and from an environmental and social perspective, an efficient use of by-products is also needed. This fact is in agreement with the circular economy policy promoted by the European Union, which strongly encourages the food supply chain to upgrade their low-quality by-products to minimize the generation of wastes (EU, 2015). These by-products can be derived from underused food products, renewable resources, and the valorization of agro-industrial and marine wastes. Husks, feathers, shells, skins, seeds, peels, stems, and leaves, among other things, are typical examples of by-products with a high valorization potential. These materials are a source of natural bioactive substances with interesting antimicrobial and/or antioxidant activities, which, in turn, could be used to develop active films/coatings for food preservation. A very well-known example of this kind of material is chitosan, a natural carbohydrate polymer, usually obtained by N-deacetylation of chitin from shrimp and crab shells, with interesting film-forming ability and antimicrobial activity. The future of these bioactive coatings and biodegradable packaging materials for food preservation is very promising. These materials should be designed to provide highly specific functional performances based on the selection of the most appropriate film-forming compounds and active ingredients, assembling them in the most effective arrangement. Moreover, if all or some of the film components come from by-products, the efficiency and process sustainability will be guaranteed.

The aim of this chapter is to find out the potential use of these by-product materials to develop coating/film packaging materials for food preservation. In the different sections, biopolymers and functional compounds derived from by-products used to develop biodegradable materials have been reviewed. The effect of the use of these by-product functional fractions when incorporated into biopolymer matrices was explored, and their benefits and drawbacks have been highlighted.

2. Biopolymers derived from plant or animal by-products

In recent years, a large number of biopolymers obtained from renewable sources have been used to develop food packaging materials. In Table 6.1, a summary of the most important biopolymers obtained from agro-industrial and marine wastes

Table 6.1 Biopolymers derived from plant or animal by-products used to obtain films by casting.

Biopolymers	Source	Main composition	Other components or/ and application	Main results	Reference
Chitosan	Crustaceous exoskeleton	D-Glucosamine and *N*-acetyl-D-glucosamine	CMC + HPMC Application: Strawberry	The coatings of CMC 1% + CH 1% and HPMC 1% + CH1% were the most effective in preserving strawberry fruit quality throughout storage.	Gol et al. (2013)
			CMC Application: Mandarins, oranges, and grape fruits	CMC/chitosan films enhanced fruit glossiness and appearance and were water resistant while maintaining their stability during exposure to cold storage and shelf live conditions.	Arnon et al. (2014)
Soy protein	Soybeans	Albumins and globulins	Cellulose nanocrystals and *Cedrus deodara* pine needle extract	Enhanced mechanical strength, antioxidant activity, and water vapor barrier capacity.	Yu et al. (2018)
			Gelatin Application: Beef patty	Coatings were effective in delaying lipid oxidation and color deterioration.	Guerrero et al. (2015)
Soy protein and zein	Soybeans Corn	Albumins and globulins Glutamic acid, leucine, proline, and alanine	–	3:1 SPI-zein films presented the best mechanical properties.	Wang et al. (2016)
Soy protein and chitosan	Soybean Crustaceous exoskeleton	Albumins and globulins-β-D-glucosamine and *N*-acetyl-D-glucosamine	$CuSO_4$ nanoclusters	Improvement of film's flexibility. Addition of $CuSO_4$ nanoclusters improved the compatibility between chitosan and soy protein isolate.	Li et al. (2017)

Continued

Table 6.1 Biopolymers derived from plant or animal by-products used to obtain films by casting.—*cont'd*

Biopolymers	Source	Main composition	Other components or/ and application	Main results	Reference
Starch	Mango kernel	Amylose and amylopectin	Application: Tomato	Delay in the development of the ripening parameters: loss weight, soluble solids, titratable acidity, ascorbic acid content, and firmness during storage at 20°C.	Nawab et al. (2017)
	Pea		Guar gum Application: Oranges	Coatings decreased the respiration rate, the production of ethylene, weight, and firmness loss, while maintaining the flavor during the fruit shelf life.	Saberi et al. (2018)
	Rice		λ-Carrageenan and stearic acid	Improvement of physical, mechanical, and barrier properties of pure starch films.	Thakur et al. (2016)
Mucilage	Prickly pears peels from *Opuntia ficus-indica* species	Galactose, arabinose, xylose, and galacturonic acid	–	Red-colored films were obtained with high water contact angle and solubility.	Gheribi et al. (2019)
Casein	Skimmed milk	Glutamic acid, leucine, glutamine, valine, etc.	Tannic acid	The addition of tannic acid to the films of casein produced cross-linking reaction between casein and tannic acid, improving the physicochemical properties.	Picchio et al. (2018)
Keratin	Chicken feathers	Glycine, cystine, cysteine	–	WVP values of the keratin films were lower than those of other protein-based films. These values were higher using sorbitol than glycerol, as plasticizer. The solubility of films increased with the addition of sorbitol.	Martelli et al. (2006)
Gelatin and defatted soy protein	Fish Soybeans	Glycine, proline, glutamic acid, albumins, globulins, etc.	Papaya puree	Gelatin increased tensile and seal strength. Defatted soy protein and gelatin increased ($P < .05$) elongation, WVP, and water contact angle and decreased water solubility values.	Tulamandi et al. (2016)

CMC, *carboxymethyl cellulose;* HPMC, *hydroxypropymethyl cellulose;* WVP, *water vapor permeability.*

used to develop packaging materials is shown. These biopolymers have excellent film-forming capacity, giving rise to edible coatings or biodegradable films with high gas barrier capacity, safe for the human health and the environment. Nevertheless, the use of these biopolymers usually presents inherent problems due to their hydrophilic nature, as they usually exhibit poor mechanical and water vapor barrier properties and/or physical deficiencies (i.e., high water solubility) in comparison with the conventional plastic materials. This is why most of the recent studies into biopolymer films have focused on the development of new strategies to obtain water-resistant films with improved mechanical and barrier properties, to meet the requirements of the food packaging materials. These strategies often involve blending different biopolymers and the incorporation of micro- or nanoparticles or additives that confer additional functionalities, which were not in the original film. The chemical modification of the polymer structure to modify its properties (i.e., making it more hydrophobic) has also been explored (Cano et al., 2016).

Among the biopolymers derived from plant by-products, proteins have been widely studied (Table 6.1). Soy protein, zein, and gluten extracted from plant residues or wastes are among the most commonly used to obtain films. Thus, soy protein can be obtained from soybean oil processing or from defatted soybean flour; gluten is a by-product from the production of wheat starch; and zein is usually extracted from corn by using suitable solvents, such as ketones, amide solvents, and esters. These compounds have excellent oxygen barrier properties, which has allowed the shelf life of beef patties treated with soy-based coatings to be extended, without affecting their sensory parameters (Guerrero et al., 2015).

To improve the mechanical properties and high moisture sensitivity of the plant protein–based films, blending of different biopolymers has been studied. In this sense, Wang et al. (2016) and Yu et al. (2018) developed improved soy protein–based films by blending with zein and cellulose nanocrystals (CNCs) or pine needle extract, respectively, giving rise to more stretchable and mechanically resistant films. Other authors improved the water, oxygen, and light barrier properties of these films by incorporating different plant extracts, thanks to the interactions between active groups of the matrix and phenolic hydroxyl groups of the extract, as commented on in the following sections.

Other biopolymers derived from plant by-products widely used to develop films are polysaccharides, such as mucilage and starch. Gheribi et al. (2019) obtained mucilage-based films with high moisture content, leading to a high water solubility and good extensibility but poor tensile strength. In comparison with other polysaccharide-based films, the prickly pear peel mucilage films showed higher water vapor barrier capacity. Thakur et al. (2016) developed rice starch films with stearic acid and ι-carrageenan, which is extracted from certain species of red seaweed of the family Rhodophyceae. It was observed that an increase in the concentration of ι-carrageenan with a low concentration of stearic acid was effective in improving the physical, mechanical, and barrier properties of the pure starch-based film.

Of the biopolymers obtained from animal by-products, those based on casein, keratin, and gelatin and chitosan are significantly noteworthy. Casein is obtained by acid precipitation or coagulation of milk. Keratin comes from chicken feathers treated with 2-mercaptoethanol in a concentrated urea solution using sodium dodecyl sulfate (Martelli et al., 2006) and gelatin and is produced by the partial hydrolysis or thermal degradation of collagen obtained from bovine skin and bones or fish skin. Likewise, chitosan can be obtained from the crustaceous exoskeleton by the deacetylation of chitin under alkaline conditions (Li et al., 2017).

Casein and chitosan-based films are transparent, flexible, and water-soluble and exhibit excellent organoleptic properties, but they are not very efficient water vapor barriers (Atarés et al., 2010). Keratin-based films have a lower water vapor permeability than other protein-based films, such as soy and gluten proteins.

Biopolymers from animal and plant by-products have also been blended to develop packaging materials with better properties. Hassan and Norziah (2012) showed that different ratios of sago starch and fish gelatin blends, plasticized with glycerol or sorbitol, affected physical, mechanical, and water vapor permeability of the films. Thus, starch/gelatin ratios of 3:1, 4:1, and 5:1 appeared to form good flexible glycerol-plasticized films, but not at higher gelatin content (2:1). Soy protein has also been combined with gelatin (Tulamandi et al., 2016) and chitosan (Li et al., 2017) to improve the mechanical properties and water resistance of soy-based films. For example, the addition of defatted soy protein and gelatin to the papaya puree produced a significant increment in the elongation, water permeability, water contact angle, and water resistance of the films (Tulamandi et al., 2016).

As can be seen in Table 6.1, these kinds of edible materials have also been used to coat different foodstuffs, such as fruits, vegetables, or nuts, as they are edible, odorless, and tasteless and have a remarkably low oxygen permeability and exert a fair barrier against water vapor. Some coatings have been shown to control the ripening process and weight loss of fruits and vegetables, providing them with greater firmness and higher total soluble solid content throughout their storage (Nawab et al., 2017; Saberi et al., 2018).

3. Active compounds from plant/animal by-products

Active compounds from food by-products can be divided according to their origin into animal and plant materials from different food industries as shown in Table 6.2: fruit and vegetable industry, grain processing, brewery and winery, dairy, marine, and meat industries. Each sector creates a large amount of different residues, such as seeds, pulps, peels, leaves, and stems from plant by-products and skin, bones, and shells from animal by-products, which contain various valuable compounds that can be extracted and reused for different purposes. The medical, cosmetic, and food industries are among the sectors most interested in using such substances.

As can be observed in Table 6.2, these active compounds represent a very heterogeneous class of chemicals, including mainly phenolic compounds (phenolic acids, flavonoids, tannins, and stilbenes), phytosterols, vitamins, and dietary fibers and biopeptides, which are mostly of plant origin.

Table 6.2 Industry by-products and their active compounds.

Industry (product)	By-product	Active compound	Activity	Reference
Fruit and vegetable industry				
Citrus fruits - Lemon, oranges, bergamot, Mandarin	Seed, peel, pulp	Phenolics	AO	Bocco et al. (1998)
Exotic fruits - Mango, jackfruit, banana, avocado	Seeds, pulp, peels	Phenolics, carotenoids, fibers	AO	Ayala-Zavala et al. (2011)
Berries - Red current, grapes	Seed	Phytosterols	AO	Górnaś and Rudzinska (2016)
Pomes - Apple, pomegranate	Seed, peel, pulp	Phytosterols, phenolics, ascorbic acid	AO, AM	Ayala-Zavala et al. (2011), Agourram et al. (2013), Górnaś and Rudzinska (2016)
Legumes - Soybean - Pea, lentil	Husk Hulls	Phenolics Phenolics	AO AO	Tyug et al. (2010) Oomah et al. (2011)
Nuts - Hazelnut - Peanut - Walnut - Chestnut - Sunflower seed	Skin/peel Skin Shell Peel, hulls Hulls	Phenolics Phenolics, flavonoids Phenolics, flavonoids Phenolics Phenolics	AO AO AM AO, AM AO AO	Agourram et al. (2013) Vijayalaxmi et al. (2015), de Camargo et al. (2017), Yuan et al. (2018) Vázquez et al. (2008)
Others - Cocoa bean - Olive - Eucalyptus - Coffee - Sunflower	Hulls Hulls, leaves, kernels, pomace Bark Cherry husk Hulls, leaves	Flavonoids Phenolics, carotenoids, Vitamin E Phenolics Phenolics, flavonoids Phenolics	AO AO AO AO AO	Arlorio et al. (2005), Okiyama et al., (2018) Roselló-Soto et al. (2015) Vázquez et al. (2008) Vijayalaxmi et al. (2015) De Leonardis et al. (2005), Menzel et al. (2019)
Vegetables - Potato - Leek	Pulp, peel Leaves	Phenolics Phenolics	AO AO	Agourram et al. (2013)
Cereal/grain production - Rice, wheat, corn	Straw, husk, germ, bran	Phenolics, Vitamin E, phytosterols	AO	Vijayalaxmi et al. (2015), Smuda et al. (2018)

Continued

Table 6.2 Industry by-products and their active compounds.—*cont'd*

Industry (product)	By-product	Active compound	Activity	Reference
Winery industry	Stems, leaves, pomace, seeds	Phenolics, flavonoids, stilbenes	AO, AM	Teixeira et al. (2014)
Brewery industry	Brewery waste water, spent grain	Phenolics, flavonoids	AO	Tatullo et al. (2016)
Marine industry	Skin, liver, bones, heads, gelatin Crustacean shellfish	Bioactive peptides Chitin/chitosan	AO AO, AM	Sila and Bougatef (2016) Hamed et al. (2016)
Meat industry	Blood, collagen	Bioactive peptides	AO, AM	Toldrá et al. (2016)
Dairy industry	Whey proteins (cheese)	Bioactive peptides	AO, AM	Rebouillat and Ortega-Requena (2015)

AM, *antimicrobial;* AO, *antioxidant.*

3.1 Active compounds from plant by-products

Natural/active extracts obtained from plant by-products have been reviewed extensively during the past two decades and have been shown to have comparable antioxidant activity as synthetic antioxidants, such as butylated hydroxyanisole (BHA) and butylated hydroxyl toluene (BHT) (Galanakis, 2012, Valdés et al., 2014). For example, phenolic acids, such as caffeic acid, found in all plants, have been shown to strongly inhibit lipid peroxidation of linoleic acid emulsion by 68.2% at 10 μg/mL, compared with BHA and BHT with 74.4% and 71.2% lipid oxidation inhibition, respectively, at 20 μg/mL (Gülçin, 2006). The fruit industry generates a wide range of by-products with high content of phenolic acids and other phenolic compounds, carotenoids, or vitamins, exhibiting antioxidant activity in vitro tests, such as DPPH radical assay, FRAP, TEAC, or ABTS assays. Phenolic compounds act as strong antioxidants by scavenging free radicals by the action of hydroxyl groups and conjugated double bonds. Besides the radical scavenging, phenolic compounds chelate transition metal ions, recombine radicals, or act as electron transfers forming stable products. Hence, plant extracts containing phenolic compounds with high antioxidant activity are excellent candidates for the development of active food packaging. Recent studies have shown the successful incorporation of such antioxidant extracts from plant by-products into diverse biopolymers to produce edible active packaging materials, as commented on in the following sections. For example, aqueous methanol extract of sunflower seed hulls was successfully incorporated into potato starch films (1–6 g/100 g starch formulation) using melt blending and compression molding, which were able to preserve their antioxidant activity, as shown by in vitro DPPH assay (Menzel et al., 2019).

Another promising source of antioxidant by-products lies within the cereal and grain production. Crops, such as wheat, corn, and rice, generate a large biomass of straw, hulls, germ, and bran, with high content of valuable biopolymers, such as hemicelluloses and phenolic compounds. Antioxidant extracts of cereal milling by-products, such as bran, germ, and husk, have shown a high total phenolic and carotenoid content with antioxidant activity (Smuda et al., 2018). Subcritical water extraction of wheat bran enabled the extraction of hemicelluloses with antioxidant activity due to the preserved phenolic (ferulic) acid substitution of the polymer (Ruthes et al., 2017).

Besides their antioxidant activity, many extracts from food by-products also exhibit antimicrobial activity against food-related microorganisms. For example, extracts of pomegranate peel (20 mg/mL) exhibited a wide range of activity against the growth of various food-related microorganisms, e.g., *Staphylococcus aureus* and *Bacillus cereus* (Agourram et al., 2013). This antibacterial activity heavily depended on the extraction solvent; methanol extracts exhibited the lowest MIC values and hence the greatest antimicrobial activity, compared with aqueous acetone and ethanol extracts. This activity is assumed to be correlated with the presence of tannins and their activity toward the inactivation of microbial adhesions, enzymes, and transport proteins. Similar results were found in another study of peel and seed extracts of avocado, inhibiting the growth of gram-positive and gram-negative bacteria (Raymond and Dykes, 2010). In that work, ethanol extracts exhibited the strongest antimicrobial activity (MIC 104–417 μg/mL) against seven bacteria and one yeast, whereas water extracts inhibited only two microorganisms.

3.2 Active compounds from animal by-products

Animal by-products are rich in proteins, essential free amino acids, group B vitamins, and bioavailable hemo-iron. The meat processing industry produces residues and by-products with high protein content, such as blood and collagen, which are an abundant source of bioactive peptides. Bioactive peptides usually consist of 5–11 amino acids, including hydrophobic amino acids, such as tyrosine or proline, which are well known for their antioxidant and antibacterial activity (Perez Espitia et al., 2012). The antioxidant action of bioactive peptides relates to the exposure of these hydrophobic and more ionizable groups through protein hydrolysis. Hence, to gain such bioactive peptides, protein-rich meat derivatives are hydrolyzed under controlled conditions with commercial peptidases followed by purification through fractionation and filtration (Toldrá et al., 2016). For example, bovine myofibrillar and connective tissue have been used successfully to produce mixtures of antioxidant peptides through hydrolysis via bacterial and fungal proteases (Ryder et al., 2016). Besides connective tissue, blood, and collagen, by-products, such as feathers, hairs, nails, and skin, contain large amounts of keratin, which can likewise be used to produce antioxidant peptides. For example, feather bioconversion by *Chryseobacterium* produced bioactive peptides with antioxidant activity, which has been evaluated using ABTS radical scavenging assay and the total reactive antioxidant potential method and has been shown to inhibit lipid oxidation (TBARS assay) (Fontoura et al., 2019).

Other sources of bioactive peptides are whey proteins, a dairy industry by-product of cheese or casein production. Whey protein hydrolysates obtained using commercial proteases have been shown to possess higher radical scavenging than synthetic BHA (Hernández-Ledesma et al., 2005). Likewise, the marine processing industry generates by-products, such as the skin, heads, bones, and livers of fish, which are also rich in protein and sources of bioactive peptides with high antioxidant activity that can be extracted as described above (Sila and Bougatef, 2016). In addition, the shells of crustacean shellfish, such as shrimps, are rich in chitin, which can be used to produce bioactive chitosan (as commented on in the previous section). Chitosan has been used together with active extracts from by-products, such as apple peel (Riaz et al., 2018), black soy bean coat (Wang et al., 2018), or blackberry and blueberry pomace (Kurek et al., 2018), to produce antioxidant and antimicrobial films.

Bioactive peptides have been successfully incorporated into edible films, retaining their bioactive properties (Perez Espitia et al., 2012). Furthermore, the protein-rich by-product from the agar–agar production of algae has been shown to be valuable for the production of antioxidant films (Cian et al., 2014). The residue during the cold water extraction from red seaweed exhibited excellent film-forming properties using solution casting, and the films showed a fourfold increase in antioxidant capacity using TEAC assay compared with their agar films, which was correlated to the high content of phenolic compounds in the extract.

Besides their antioxidant activity, many bioactive peptides from animal by-products exhibit antimicrobial activity, e.g., bovine hemoglobin hydrolysates extracted using pig pepsin inhibited the bacterial growth of, e.g., *Listeria innocua*, *E. coli*, and *S. aureus* (MIC 1–187.1 μM) using microtiter plate assay (Adje et al., 2011). In addition, bioactive peptides from whey proteins using trypsin hydrolysis have exhibited antibacterial activity against *Listeria monocytogenes* and *Staphylococcus aureus* (MIC 10–20 mg/mL) (Demers-Mathieu et al., 2013). However, to the best of our knowledge, little research has been conducted into the antimicrobial activity of active peptides from by-products in the production of edible films in food packaging.

4. Reinforcing agents from plant/animal by-products

An alternative means to improving the functional properties of packaging materials for food applications is through the incorporation of micro- or nanofillers into the matrix for the purposes of obtaining micro- or nanocomposites. Composites are made by immiscible phases constituted by the polymer continuous network in which the filler particles are dispersed. The filler can positively modify the functional characteristics of the material, depending on the filler-polymer network interactions (Ortega-Toro et al., 2017). These interactions are often determined by the size (micro or nano) and the final filler content in the composite. Both the presence of fine dispersions homogeneously dispersed within the matrix and the establishment of a good interfacial adhesion between the fillers and the continuous matrix are usually required.

Table 6.3 summarizes the commonly used fillers, at micro- and nanoscale, obtained from plant or animal by-products, incorporated into different polymeric matrices. The main characteristics of the filler are shown, as well as the filler content in the composite and processing method. The improvement in the mechanical properties of composites brought about the presence of filler agents is mainly due to the addition of a hard phase, which absorbs part of the external stress (Ferreira et al., 2018). In addition, the barrier properties could be enhanced because the filler acts as a blocking agent within the polymeric matrix, promoting a tortuous path to the permeation of the water and gas molecules. Also, an improvement in biodegradation has been reported, mainly due to the hydrophilic nature of cellulose fibers or CNCs.

As concerns the fillers from plant by-products, the most widely used materials are cellulose or cellulose derivative fibers. Lignocellulosic materials are one of the most attractive natural sources of renewable fillers because of their biodegradability, low density, and excellent mechanical properties. Some authors have tried to incorporate these fibers directly into matrices by mechanical grinding, the obtained results being negatively affected by their large particle size and poor interfacial adhesion between the fiber/matrix, which promoted the formation of voids and holes around the fibers, thus resulting in poor mechanical behavior (Berthet et al., 2015).

To avoid the aforementioned problems, the use of fibers, which are smaller, more uniform, and compatible with the matrix, has been further explored. Microcellulose fibers or CNCs have been widely isolated from lignocellulosic materials through some different processes, which involve mechanical pretreatment, alkali treatment with subsequent bleaching process, and acid hydrolysis, as shown in Table 6.4. CNCs have also been successfully isolated from rice and coffee husks, marine, and starch-rich wastes, as described by several authors. Collazo-Bigliardi et al. (2018) isolated cellulosic fractions from rice and coffee husks to be incorporated into starch matrices. All of the functional properties of the films were improved, especially after the addition of nanoscale fillers. Similar results were described by Sung et al. (2017) and Fortunati et al. (2015) when working on PLA blends with CNC from coffee silverskin and red *Posidonia oceanica* wastes, respectively. Mukurubira et al. (2017) and González et al. (2015) incorporated starch nanocrystals (SNCs) into corn and potato thermoplastic starch matrices, respectively. The inclusion of SNC significantly decreased ($P < .05$) the water vapor permeance and oxygen permeability, thus indicating the effectiveness of hydrogen bonding at the interphase. The mechanical properties were improved while the thermal stability and opacity were increased.

Table 6.3 Reinforcing agents from plant/animal by-products or natural resources.

Resource	Filler	Extraction method/final product	Properties				Composite (polymer/filler content/ processing method)	References
			Chemical composition	Thermal analysis	Degree of crystallinity	Aspect ratio (L/d)		
Microreinforcing agents								
Plant by-products								
Wheat straw grains	Lignocellulosic fibers	Cut milling, grinding/dried fibers	Hemicellulose: 20.5% Cellulose: 32% Lignin: 17.4%	T_{Peak}: 315–342°C	–	1–100	PHBV/10, 20, 30, 40, 50 wt%/ extrusion and compression molding	Berthet et al. (2015)
Brewing spent grains	Lignocellulosic fibers	Washed in water, sieving, 3% NaOH 70°C for 20 min and 6% NaOH, 90°C for 20 min, grinding/dried fibers	Hemicellulose: 7% Cellulose: 55.6% Lignin:16.2%	T_{Peak}: 315–342°C	–	1–100	PHBV/10, 20, 30, 40, 50 wt%/ extrusion and compression molding	
Olive pomace	Lignocellulosic fibers	Cryocentrifugal milling, phenolic extraction, residual oil removal, proteins chemical removal, grinding/dried fibers	Hemicellulose: 10% Cellulose: 38% Lignin: 49%	T_{Peak}: 315–342°C	–	1–100	PHBV/10, 20, 30, 40, 50 wt%/ extrusion and compression molding	

Coffee husk	Cellulose fibers	Alkali treatment: 4 wt% of NaOH at 80°C. Bleaching treatment: acetate buffer, 1.7 wt% sodium chlorite and distilled water at reflux temperature/ dried fibers.	Hemicellulose: 27.2% Cellulose: 61.8% Lignin: 2.6%	T_{Onset}: 280°C T_{Peak}: 316-°C	52%	L:70–570 μm	Thermoplastic corn starch/1, 5, 10 wt%/melt blending and compression molding	Collazo-Bigliardi et al. (2018)
Rice husk	Cellulose fibers		Hemicellulose: 19.2% Cellulose: 73.8% Lignin: 1.6%	T_{Onset}: 292°C T_{Peak}: 327°C	50%	L:60–490 μm	Thermoplastic corn starch/1, 5, 10 wt%/melt blending and compression molding	
Almond shell	Cellulose fibers	Sieving, drying, 5 w/vt% NaOH 80°C, 10 v/v% H_2O_2 80°C, grinding, drying/ dried fibers	*Raw material Hemicellulose: 7.56% Cellulose: 46.25% Lignin: 29.85%	T_{Peak}: ~ 340°C	57%	2	PHB/10, 20 wt%/internal mixer and compression molding	Sánchez-Safont et al. (2018)
Rice husk	Cellulose fibers	Ultracentrifugal milling, drying, 5 w/vt% NaOH 80°C, 10 v/v% H_2O_2 80°C, grinding, drying/ dried fibers	*Raw material Hemicellulose: 18.6% Cellulose: 31.13% Lignin: 28.25%	T_{Peak}: ~ 350°C	66%	7		
Seagrass	Cellulose fibers	Ultracentrifugal milling, drying, 3 v/v% CH_3COOH, 5 w/vt% NaOH 80°C, 10 v/v% H_2O_2 80°C, grinding, drying/ dried fibers	*Raw material Hemicellulose: 21% Cellulose: 38% Lignin: 27%	T_{Peak}: ~ 330°C	50%	2–15		

Continued

Table 6.3 Reinforcing agents from plant/animal by-products or natural resources.—*cont'd*

Resource	Filler	Extraction method/final product	Properties				Composite (polymer/filler content/ processing method)	References
			Chemical composition	Thermal analysis	Degree of crystallinity	Aspect ratio (L/d)		
Animal by-products								
Poultry feathers	Keratin	Sanitized with ethanol, died, ground in a rotational blade and milling with high impact and shear stress to the material/ dried powder	Protein: cysteine	T_{Peak}: 265–350°C	–	6	PHBV/0.5, 1, 3, 5, 10, 25, 50 WT%/internal mixer and compression molding	Pardo-Ibáñez et al. (2014)
Bovine hair skin	Keratin fibers	Rinsed, degreased, dried (outdoor and oven at 40°C for 12 h), milled/dried fibers	Protein: 91%	T_{Peak}: 340°C	–	d: 16–83 μm	PLA/3, 5, 10 wt%/extrusion and injection molding	Sanchez-Olivares et al. (2017)
Nanoreinforcing agents								
Plant by-products								
Coffee silverskin	Cellulose nanocrystals	Grinding Alkali treatment: 5 wt/v of KOH at 80°C. Bleaching treatment: acetate buffer, 1.5 w/v% sodium hypochlorite at 70°C. Acid hydrolysis: 64% H_2SO_4 at 45°C/ freeze-dried powder	Raw material Hemicellulose: 17% Cellulose: 24% Lignin: 29%	–	72%	9.4–10.5	PLA/1, 3, 5 wt%/extrusion and compression molding	Sung et al. (2017)

Coffee husk	Cellulose nanocrystals	Alkali treatment: 4 wt% of NaOH at 80°C. Bleaching treatment: acetate buffer, 1.7wt% sodium chlorite and distilled water at reflux temperature. Acid hydrolysis: 64% H_2SO_4 at 45°C/Water dispersion	Raw material Hemicellulose: 18.2% Cellulose: 35.4% Lignin: 23.2%	T_{Onset}: 249°C T_{Peak}: 296°C	92%	10–40	Thermoplastic corn starch/ 1 wt%/melt blending and compression molding	Collazo-Bigliardi et al. (2018)
Rice husk			Raw material Hemicellulose: 17.1% Cellulose: 33.8% Lignin: 21.5%	T_{Onset}: 262°C T_{Peak}: 301°C	90%	10–40		
Posidonia oceanica waste	Cellulose nanocrystals	64% H_2SO_4 at 45°C. Addition of surfactant, (Beycostat A B09-CECCA S.A.)/freeze-dried powder	Raw material Cellulose: 40%	T_{Peak}: 291°C	-	34.6	PLA/1, 3 wt%/ casting	Fortunati et al. (2015)
Amadumbe corms (*Colocasia esculenta*)	Starch nanocrystals	H_2SO_4 (3.16M) at 40°C for 5 days. Centrifugation, sonication/ freeze-dried powder	Amylose: 7%–8%	T_{Peak}: 126°C	–	d: 5–100 nm	Potato starch/ 2.5, 5, 10 wt%/ casting	Mukurubira et al. (2017)
Waxy maize starch	Starch nanocrystals	H_2SO_4 (3.16M) at 40°C for 5 days. Centrifugation, sonication/water dispersion	Amylose: 25%	–	22.8%	1.2–1.3	Thermoplastic corn starch/1, 2.5, 5 wt%/ casting	González et al. (2015)

Continued

Table 6.3 Reinforcing agents from plant/animal by-products or natural resources.—*cont'd*

Resource	Filler	Extraction method/final product	Properties				Composite (polymer/filler content/ processing method)	References
			Chemical composition	Thermal analysis	Degree of crystallinity	Aspect ratio (L/d)		
Animal by-products								
Tunicate (*Ciona intestinalis*)	Cellulose nanocrystals (CNs)	[1]Enzymatic hydrolysis [2]TEMPO-mediated oxidation [3]Acid hydrolysis: 55 H_2SO_4 at 60°C	–	T_{Peak} [1]260°C, [2]20°C, [3]120°C	[1]82.8% [2]87.5% [3]90.7%	[1]d: 17.1 nm [2]59 −131 [3]22–44	Glucomannans (G)/80:20, 50: 50, CN:G/ Casting	Zhao et al. (2015)
Lobster waste	Chitin nanocrystals	2 M NaOH solution for 24 h at 25°C, HCl (2M) for 3 h at 25°C, acetone, ethanol. Acid hydrolysis 3 M HCl at 100°C for 90 min/water dispersion	Polysaccharide consisting of β(1−4)-linked *N*-acetyl-D-glucosamine units	T_{Onset}: 242°C T_{Peak}: 299°C	89%	5	Thermoplastic corn starch/2, 5, 10, 15, 20 wt%/melt blending and compression molding	Salaberria et al. (2014)
Crab shell	Chitin nanofibrils	Acid hydrolysis 3N HCl at 60°C for 3 h, centrifuged, washed, homogenized, dialyzed/water dispersion	Linear polysaccharide, made up of β-(1−4)-linked 2-acetamido-2-deoxy-D-glucopyranose units	T_{Peak}: 381°C	88.6%	20–30	Carrageenan/3, 5, 10 wt%/ casting	Shankar et al. (2015)

Table 6.4 Active biopolymer films in food applications by incorporating active fractions from by-products or natural resources.

Biopolymers	Active product	Application	Main results	References
From plants				
Zein	Pomegranate peel extract	Himalayan cheese	Delay in the oxidation reactions and microbial spoilage during storage.	Mushtaq et al. (2018)
Gluten	Gallic and tannic acid, flaxseed gum, oligomeric procyanidins, and lauric acid	–	Improvement in the mechanical and barrier properties.	Hager et al. (2012) Liu et al. (2018)
From animals				
Chitosan	Green tea extract	Pork sausages	Delay in of the lipid oxidation and microbial growth and extending shelf life, while maintaining organoleptic qualities of the sausages.	Siripatrawan and Noipha (2012)
Chitosan	–	Meat burger	Better preservation of the microbial quality.	Vargas et al. (2011)
Soy protein isolate	Thyme, oregano essential oils	Ground beef	*Escherichia coli, E. coli* O157:H7, and *S. aureus* were significantly inhibited by antimicrobial films, *Lactobacillus plantarum* and *Pseudomonas aeruginosa* appeared to be the more resistant bacteria.	Emiroğlu et al. (2010)
Keratin	Gelatin and clove oil	Smoked salmon	Antimicrobial properties against *E. coli* O157:H7 and *Listeria monocytogenes*.	Song et al. (2014)

Continued

Table 6.4 Active biopolymer films in food applications by incorporating active fractions from by-products or natural resources.—*cont'd*

Biopolymers	Active product	Application	Main results	References
Chitosan/gelatin	MC and tannic acid	Cherry tomatoes and grapes	Lower weight loss and browning index values than samples coated with control films.	Halim et al. (2018)
Chitosan/gelatin	–	Red bells peppers	Inhibition of microbial decay. Extended shelf life up to 3 weeks under cold-storage conditions and up to 14 days at ambient temperature.	Poverenov et al. (2014)
Chitosan/gelatin	Clove essential oil	Salmon carpaccio	Retardation of the growth of microorganisms (*Pseudomonas*). Inhibition of Enterobacteriaceae and lipid oxidation.	Gómez-Estaca et al. (2018)
Casein	Sodium alginate-pectin and Cu^{+2} nanocomposites	Coconut oil	Prolongation of the shelf life of coconut oil, decreasing the lipid oxidation rate during storage.	Gautam and Mishra (2016)
From plants and animals				
Chitosan/quinoa protein	–	Strawberry	Lower amount of mold and yeast growth than the uncoated strawberries. Retention of total sensorial quality.	Valenzuela et al. (2015)
Chitosan/HPMC	Bergamot essential oil	Table grapes	Improvement of weight losses, fruit firmness, slowed down the respiration rates and antimicrobial activity.	Sánchez-González et al. (2011)
Chitosan/CMC/casein	Rosemary, oregano, olive, capsicum, garlic, onion, cranberry, oreganum, and oreganum + carvacrol 5%	Squash slices	Antioxidant protection of the minimally processed squash.	Ponce at al. (2008)

MC, *methylcellulose.*

As regard fillers from animal by-products, keratin fibers (KFs), which are extracted mainly from poultry feathers and bovine hair skin at microscale, are also widely used. Their incorporation into films has been shown to improve the mechanical resistance of composites, as their elastics modulus, tensile strength, flexural modulus, and hardness significantly increased. KFs have also been reported to increase the polymer thermal stability better than cellulosic fibers (Baba and Özmen, 2015). Thus, Sanchez-Olivares et al. (2017) incorporated 3, 5, and 10 wt% of keratin fibers derived from bovine hair skin into PLA blends with a flame retardant, by injection molding. The formulations with higher KF contents exhibited better flow properties, making the PLA processing easier. Moreover, a good KF/polymer matrix adhesion was observed, which led to a 16% increase in tensile strength, 40% in strain at break, and 66% in tenacity. Similar results were found by Pardo-Ibáñez et al. (2014) with the addition of 0.5–50 wt% of KF in PHBV-based composites obtained by melt blending and compression molding. These results are explained because of the good physical interactions established between the polymeric matrix and the filler, as observed by SEM analysis.

As regard the nanoscale fillers from animal by-products, chitin nanocrystals/nanofibrils have been mostly isolated from crustacean waste products (lobster waste, crab shell, prawn shell) as described by Shankar et al. (2015). These authors developedchitin nanofibrils (CNFs)–reinforced carrageenan nanocomposite films by using the solution-casting technique. The improvement in their mechanical properties can be associated with the good dispersion of the nanosize fillers in the matrix, resulting from their chemical similarity, and also with the strong nanofiller-matrix adhesion by hydrogen bonding interactions. Some authors also relate the reinforcing effect of the chitin nanocrystals to their high degree of stiffness and strength, and the increase in the mechanical properties of the composite is correlated to the formation of a percolating network based on hydrogen bonding forces. In contrast, the elongation at break decreased linearly with the increase in the filler content caused by the rigid nature of these nanofillers (Barhordari and Fathi, 2018). Their water vapor permeability and transparency decreased slightly as did the elongation at break due to CNF reinforcement; however, the tensile strength and modulus of carrageenan film increased after CNF incorporation up to 5 wt%. This behavior was also observed by Salaberria et al. (2014) after the addition of chitin nanocrystals to thermoplastic starch blends by injection molding.

5. Combination of biopolymers and compounds obtained from by-products to obtain improved food packaging materials

In recent years, a large number of biopolymers obtained from renewable sources have been used in combination with by-product fractions from plants and animals for the development of improved food packaging materials, as can be observed in

Table 6.4. These packaging materials have been applied to different foodstuffs, such as fruits, red peppers, cheese, meat, coconut oil, fish, and bread, with the aim of limiting food deterioration and increasing its shelf life. Within the active compounds added to biopolymer films, phenolic compounds, tannins, essential oils, and antimicrobial nanoparticles and biopolymers possessing antimicrobial activity per se have been found to generate wider interest among researchers (Table 6.4). For example, gluten has been blended with gallic and tannic acid (Hager et al., 2012) as well as with flaxseed gum, oligomeric procyanidins, and lauric acid (Liu et al., 2018). The results depended on the interactions established between polymer matrix and the active compound. Thus, films containing tannic acid became stiffer and more resistant but less stretchable due to a cross-linking effect. On the other hand, films containing gallic acid became more elastic because gallic acid acted as a plasticizer. The oligomeric procyanidins also had a cross-linking effect on the gluten proteins and reduced the free volume of the polymer matrix, thus leading to the formation of a more compact film structure, which contributes to a reduction in the water vapor permeability of the film (Nie et al., 2015).

In addition to the enhancement of the physical properties, most of the active agents incorporated into film matrices exhibited antimicrobial or antioxidant activity that can be conferred to the film. The extracts of the pomegranate peel incorporated into zein films (2.5−7.5 g/100 g zein) has been demonstrated to exhibit dose-dependent antimicrobial activity against *E. coli, Clostridium perfringens, Micrococcus luteus, Enterococcus faecalis, S. aureus, Proteus vulgaris,* and *Salmonella typhii* microbial, which was attributed to the polyphenol and tannin content in the extract (Mushtaq et al., 2018). This study showed that these antimicrobial zein films protected Himalaya cheese against spoilage, seen as a decrease in the total bacterial, mold, and yeast counts up to 15 days' storage. Aqueous extracts of grape seeds from winery production incorporated into chitosan films also showed antimicrobial activity against *S. aureus, B. cereus, Bacillus subtilis*, and *E. coli* in in vitro studies (Sogut and Seydim, 2018). These antimicrobial films were applied to chicken breast fillets, thus significantly reducing the total mesophilic aerobic and coliform bacteria of the product after 15 storage days and promoting the inhibition of the lipid oxidation process. In other studies, grape seed, green tea, and tea polyphenol extracts incorporated into pea starch or chitosan films have been shown to inhibit the lipid oxidation and microbial growth of the meat products, while maintaining the quality and increasing their shelf life (Corrales et al., 2009; Siripatrawan and Noipha, 2012). Gelatin films incorporating dried and milled beet root residues (Iahnke et al., 2016) also provided protection against the lipid oxidation of sunflower oil, showing peroxide values under 10 mEq/kg after 35 days in extreme conditions (35°C at 54% relative humidity exposed to fluorescent light at >900 lux).

Some biopolymers, such as chitosan, exhibit antimicrobial activity and have been used as an active material for the packaging of strawberries, mandarins, oranges, and grapes (Gol et al., 2013; Arnon et al., 2014) among others. Edible coatings based on chitosan have the ability both to act as a gas barrier by modifying the internal atmosphere of the fruit and also to provide antimicrobial activity. The exact mechanism of the antimicrobial action of chitosan is not well known although it is believed to be related with its amino groups, which show a positive charge and can interact with the cell wall, changing its permeability, whereby the cytoplasma flows out, causing membrane leakage (Cuero, 1999). Several authors have coated strawberries with chitosan-based films, thus producing a beneficial impact in maintaining their quality and preventing fungal deterioration during postharvest handling (Gol et al., 2013; Vargas et al., 2006). Edible films or coatings based on chitosan and gelatin, with the addition of other compounds, have also been applied on cherry tomatoes, grapes, and red bell peppers, reducing the microbial growth, prolonging the shelf life, and improving the stability of the coated samples (Poverenov et al., 2014; Halim et al., 2018). Vargas et al. (2011) developed chitosan films to be applied on the surface of pork burgers. The results showed that chitosan-based films increased the metmyoglobin content of coated hamburgers during cold storage. Furthermore, the addition of sunflower oil to chitosan films gave rise to brighter films with improved water vapor barrier properties.

Essential oils and their components, commonly used as flavoring agents in the food industry, also present interesting antibacterial, antifungal, and antioxidant properties, which can be conferred to the films. Even if their mechanisms of action are not clearly described, it seems that the antimicrobial activity is essentially due to their hydrophobicity. Terpenes, the major compounds of essential oil, have the ability to disrupt and penetrate the lipid structure of the bacteria cell membrane, leading to the denaturing of proteins and the destruction of the cell membrane (Turina et al., 2006). Essential oils include terpenoid and phenolic compounds, which also present notable antioxidant activity. Thus, chitosan, whey protein isolate, and gelatin films blended with different essential oils have been shown to provide remarkable antimicrobial and antioxidant activities when applied mainly to fruits, meat, and fish products (Table 6.4). Thus, Gómez-Estaca et al. (2018) and Song et al. (2014) developed an antimicrobial film based on clove essential oil to slow down the growth of microorganisms (Pseudomonas) and the lipid oxidation process of salmon slices.

Gautman et al. (2016) developed a bilayer of alginate-pectin/sodium caseinate incorporating copper nanocomposites to enhance the physical properties of the films while providing antioxidant activity. Thus, the addition of these nanoparticles improved the mechanical and barrier properties of the films, lengthening the shelf life of coconut oil by decreasing its lipid oxidation rate during storage.

It is worth commenting that, in most of the studies, there is a lack of correlation between the antimicrobial results obtained in vitro (e.g., using agar diffusion methods) and in vivo studies using real food matrices. This has been attributed both to the interactions between the active substance and food components and to the limitations in the release of the active agents from the films to the food surface. Therefore, studies into real foods are necessary to validate the active properties of the developed materials.

6. Conclusions

From an economic and environmental point of view, efficient reutilization of by-products is necessary. This is in line with the vision of the circular economy that encourages, for example, the food supply chain to upgrade their low-quality by-products. It is true that this concept holds strong in theory; its translation into practice is still complicated. Remarkable efforts have been made by many researches around the world to facilitate and optimize this valorization throughout numerous scientific works. The results obtained when incorporating agri-food by-products into edible/active films to be used as food packaging materials are very promising and it could be employed as a strategy to contribute to the development of more sustainable and competitive products.

References

Adje, E.Y., Balti, R., Kouach, M., Dhulster, P., Guillochon, D., Nedjar-Arroume, N., 2011. Obtaining antimicrobial peptides by controlled peptic hydrolysis of bovine hemoglobin. International Journal of Biological Macromolecules 49 (2), 143–153.

Agourram, A., Ghirardello, D., Rantsiou, K., Zeppa, G., Belviso, S., Romane, A., Oufdou, K., Giordano, M., 2013. Phenolic content, antioxidant potential, and antimicrobial activities of fruit and vegetable by-product extracts. International Journal of Food Properties 16 (5), 1092–1104.

Arlorio, M., Coïsson, J.D., Travaglia, F., Varsaldi, F., Miglio, G., Lombardi, G., Martelli, A., 2005. Antioxidant and biological activity of phenolic pigments from theobroma cacao hulls extracted with supercritical CO_2. Food Research International 38 (89), 1009–1014.

Arnon, H., Zaitsev, Y., Porat, R., Poverenov, E., 2014. Effects of carboxymethyl cellulose and chitosan quality of citrus fruit. Postharvest Biology and Technology 87, 21–26.

Atarés, L., Bonilla, J., Chiralt, A., 2010. Characterization of sodium caseinate-based edible films incorporated with cinnamon or ginger essential oils. Journal of Food Engineering 100, 678–687.

Ayala-Zavala, J.F., Vega-Vega, V., Rosas-Domínguez, C., Palafox-Carlos, H., Villa-Rodriguez, J.A., Siddiqui, M.W., Dávila-Aviña, J.E., González-Aguilar, G.A., 2011. Agro-industrial potential of exotic fruit byproducts as a source of food additives. Food Research International 44 (7), 1866–1874.

Baba, O., Özmen, U., 2015. Preparation and mechanical characterization of chicken feather/PLA composites. Polymer Composites 38 (5), 837–845.

Barhordari, M.R., Fathi, M., 2018. Production and characterization of chitin nanocrystals from prawn shell and their application for stabilization of pickering emulsions. Food Hydrocolloids 82, 338–345.

Berthet, M.A., Angellier-Coussy, H., Machado, D., Hilliou, L., Stabler, A., Vicente, A., Gontard, N., 2015. Exploring the potentialities of using lignocellulosic fibres derived from three food by-products as constituents of biocomposites for food packaging. Industrial Crops and Products 69, 110–122.

Bocco, A., Cuvelier, M.-E., Richard, H., Berset, C., 1998. Antioxidant activity and phenolic composition of citrus peel and seed extracts. Journal of Agricultural and Food Chemistry 46 (6), 2123–2129.

Cano, A., Chafer, M., Chiralt, A., Gonzalez-Martinez, C., 2016. Strategies to improve the functionality of starch based films. In: Thakur, V.K., Thakur, M.K., Kessler, M.R. (Eds.), Handbook of Composites from Renewable Materials, Volume 4, Functionalization. Wiley, pp. 311–329.

Cian, R.E., Salgado, P.R., Drago, S.R., González, R.J., Mauri, A.N., 2014. Development of naturally activated edible films with antioxidant properties prepared from red seaweed Porphyra columbina biopolymers. Food Chemistry 146, 6–14.

Collazo-Bigliardi, S., Ortega-Toro, R., Chiralt, A., 2018. Isolation and characterisation of microcrystalline cellulose and cellulose nanocrystals from coffee husk and comparative study with rice husk. Carbohydrate Polymers 191, 205–215.

Corrales, M., Han, H.J., Tauscher, B., 2009. Antimicrobial properties of grape seed extracts and their effectiveness after incorporation into pea starch films. International Journal of Food Science and Technology 44 (2), 425–433.

Cuero, R.G., 1999. Antimicrobial action of exogenous chitosan. In: Jollés, P., Muzzarelli, R.A.A. (Eds.), Chitin and Chitinases. Birkhäuser Verlag, Basel, Switzerland, pp. 315–333.

de Camargo, A.C., Regitano-d'Arce, M.A.B., Rasera, G.B., Canniatti-Brazaca, S.G., do Prado-Silva, L., Alvarenga, V.O., Sant'Ana, A.S., Shahidi, F., 2017. Phenolic acids and flavonoids of peanut by-products: antioxidant capacity and antimicrobial effects. Food Chemistry 237, 538–544.

Demers-Mathieu, V., Gauthier, S.F., Britten, M., Fliss, I., Robitaille, G., Jean, J., 2013. Antibacterial activity of peptides extracted from tryptic hydrolyzate of whey protein by nanofiltration. International Dairy Journal 28 (2), 94–101.

De Leonardis, A., Macciola, V., Di Domenico, N., 2005. A first pilot study to produce a food antioxidant from sunflower seed shells (Helianthus annuus). European Journal of Lipid Science and Technology 107 (4), 220–227.

Emiroğlu, Z.K., Yemis, G.P., Coskun, B.K., Candogan, K., 2010. Antimicrobial activity of soy edible films incorporated with thyme and oregano essential oil on fresh ground beef patties. Meat Science 86 (2), 283–288.

European Commision, 2015. Closing the loop-An EU action for the circular economy. In: Communication from the Commission to the European Parliament, the Council, the European Economic and Social Committee and the Committee of the Regions. COM(2015) 614 final.

Ferreira, F.V., Dufresne, A., Pinheiro, I.F., Souza, D.H.S., Gouveia, R.F., Mei, L.H.I., Lona, L.M.F., 2018. How do cellulose nanocrystals affect the overall properties of biodegradable polymer nanocomposites: a comprehensive review. European Polymer Journal 108, 274–285.

Fontoura, R., Daroit, D.J., Corrêa, A.P.F., Moresco, K.S., Santi, L., Beys-da-Silva, W.O., Yates III, J.R., Moreira, J.C.F., Brandelli, A., 2019. Characterization of a novel antioxidant peptide from feather keratin hydrolysates. New Biotechnology 49, 71–76.

Fortunati, E., Luzi, F., Puglia, D., et al., 2015. Processing of PLA nanocomposites with cellulose nanocrystals extracted from Posidonia oceanica waste: innovative reuse of coastal plant. Industrial Crops and Products 67, 439–447.

Galanakis, C.M., 2012. Recovery of high added-value components from food wastes: conventional, emerging technologies and commercialized applications. Trends in Food Science & Technology 26 (2), 68–87.

Gautam, G., Mishra, P., 2016. Development and characterization of copper nanocomposites containing bilayer film for coconut oil packaging. Journal of Food Processing and Preservation 41, 132–143.

Gheribi, R., Habibi, Y., Khwaldia, 2019. Prickly pear peels as a valuable resource of added-value polysaccharide: study of structural, functional and film forming properties. International Journal of Biological Macromolecules 126, 238–245.

Gol, N.B., Patel, P.R., Ramana Rao, T.V., 2013. Improvement of quality and shelf-life of strawberries with edible coatings enriched with chitosan. Postharvest Biology and Technology 85, 185–195.

Gómez-Estaca, J., López-Caballero, M.E., Martínez-Bartolomé, M.A., López de Lacey, A.M., Gómez-Guillén, M.C., Montero, M.P., 2018. The effect of the combined use of high pressure treatment and antimicrobial edible film on the quality of salmon carpaccio. International Journal of Food Microbiology 283, 28–36.

González, K., Retegi, A., González, A., Eceiza, A., Gabilondo, N., 2015. Starch and cellulose nanocrystals together into thermoplastic starch bionanocomposites. Carbohydrate Polymers 117, 83–90.

Górnaś, P., Rudzińska, M., 2016. Seeds recovered from industry by-products of nine fruit species with a high potential utility as a source of unconventional oil for biodiesel and cosmetic and pharmaceutical sectors. Industrial Crops and Products 83, 329–338.

Guerrero, P., O'Sullivan, M.G., Kerry, J.P., De la Caba, K., 2015. Application of soy protein coatings and their effect on the quality and shelf-life stability of beef patties. Royal Society of Chemistry Advances 5, 81–82.

Gülçin, İ., 2006. Antioxidant activity of caffeic acid (3,4-dihydroxycinnamic acid). Toxicology 217 (2), 213–220.

Hager, A.,S., Vallons, K., Arendt, E., 2012. Influence of gallic acid and tannic acid on the mechanical and barrier properties of wheat gluten films. Journal of Agricultural and Food Chemistry 60, 6157–6163.

Halim, A.L., Kamari, A., Phillip, E., 2018. Chitosan, gelatin and methylcellulose films incorporated with tannic acid for packaging. International Journal of Biological Macromolecules 120, 1119–1126.

Hamed, I., Özogul, F., Regenstein, J.M., 2016. Industrial applications of crustacean by-products (chitin, chitosan, and chitooligosaccharides): a review. Trends in Food Science & Technology 48, 40–50.

Hassan, A.A., Norziah, M.H., 2012. Starch-gelatin edible films: water vapor permeability and mechanical properties as affected by plasticizers. Food Hydrocolloids 26, 108–117.

Hernández-Ledesma, B., Dávalos, A., Bartolomé, B., Amigo, L., 2005. Preparation of antioxidant enzymatic hydrolysates from α-lactalbumin and β-lactoglobulin. Identification of active peptides by HPLC-MS/MS. Journal of Agricultural and Food Chemistry 53 (3), 588–593.

Iahnke, A.O.eS., Costa, T.M.H., de Oliveira Rios, A., Flôres, S.H., 2016. Antioxidant films based on gelatin capsules and minimally processed beet root (Beta vulgaris L. Var. Conditiva) residues. Journal of Applied Polymer Science 133 (10), 43094. https://doi.org/10.1002/app.43094.

Li, K., Jin, S., Liu, X., Chen, H., He, J., Li, J., 2017. Preparation and characterization of chitosan/soy protein isolate nanocomposite film reinforced by Cu nanoclusters. Polymers 9, 247.

Liu, R., Cong, X., Song, Y., Wu, T., Zhang, M., 2018. Edible gum—phenolic—lipid incorporated gluten films for food packaging. Journal of Food Science 83, 6.

Martelli, S.M., Rocha, G., Borges, J., 2006. Mechanical properties, water vapor permeability and water affinity of feather keratin films plasticized with sorbitol. Journal of Polymers and the Environment 14, 215—222.

Menzel, C., González-Martínez, C., Chiralt, A., Vilaplana, F., 2019. Antioxidant starch films containing sunflower hull extracts. Carbohydrate Polymers 214, 142—151.

Mukurubira, A.R., Mellem, J.M., Amonsou, E.O., 2017. Effects of amadumbe starch nanocrystals on the physicochemical properties of starch biocomposite films. Carbohydrate Polymers 165, 142—148.

Mushtaq, M., Gani, A., Gani, A., Punoo, H.A., Masoodi, F.A., 2018. Use of pomegranate peel extract incorporated zein film with improved properties for prolonged shelf life of fresh Himalayan cheese (Kalari/kradi). Innovative Food Science & Emerging Technologies 48, 25—32.

Nawab, A., Alam, F., Hasnain, A., 2017. Mango kernel starch as a novel edible coating for enhancing shelf-life of tomato (Solanum lycopersicum) fruit. International Journal of Biological Macromolecules 103, 581—586.

Nie, X., Gong, Y., Wang, N., Meng, X., 2015. Preparation and characterization of edible miofibrilar protein-based film incorporated with grape seed procyanidins and green tea polyphenol. Food Science and Technology 64, 1042—1046.

Okiyama, D.C.G., Soares, I.D., Cuevas, M.S., Crevelin, E.J., Moraes, L.A.B., Melo, M.P., Oliveira, A.L., Rodrigues, C.E.C., 2018. Pressurized liquid extraction of flavanols and alkaloids from cocoa bean shell using ethanol as solvent. Food Research International 114, 20—29.

Oomah, B.D., Caspar, F., Malcolmson, L.J., Bellido, A.-S., 2011. Phenolics and antioxidant activity of lentil and pea hulls. Food Research International 44 (1), 436—441.

Ortega-Toro, R., Bonilla, J., Talens, P., Chiralt, A., 2017. Starch-based materials in food packaging. In: Villar, M.A., Barbosa, S.E., García, M.A., Castillo, L.A., López, O.V. (Eds.), Future of Starch-Based Materials in Food Packaging. Aspen Publishers, pp. 257—312.

Pardo-Ibáñez, P., Lopez-Rubio, A., Martínez-Sanz, M., Cabedo, L., Lagaron, J.M., 2014. Keratin-polyhydroxyalkanoate melt-compounded composites with improved barrier properties of interest in food packaging applications. Journal of Applied Polymer Science 39947, 1—10.

Perez Espitia, P.J., de Fátima Ferreira Soares, N., dos Reis Coimbra, J.S., de Andrade, N.J., Souza Cruz, R., Alves Medeiros, E.A., 2012. Bioactive peptides: synthesis, properties, and applications in the packaging and preservation of food. Comprehensive Reviews in Food Science and Food Safety 11 (2), 187—204.

Picchio, M.L., Garro, Y., Monti, G.A., Gugliotta, L.L., Minari, R.J., Alvarez, C.I., 2018. Casein films crosslinked by tannic acid for packaging applications. Food Hydrocolloids 84, 424—434.

Ponce, A.G., Roura, S.I., Del Valle, C.E., Moreira, M.R., 2008. Antimicrobial and antioxidant activities of edible coatings enriched with natural plant extracts: in vitro and in vivo studies. Postharvest Biology and Technology 49, 294—300.

Poverenov, E., Zaitsev, Y., Arnon, H., Granit, R., Alkalai-Tuvia, S., Perzelan, Y., Weinberg, T., Fallik, E., 2014. Effects of a composite chitosan-gelatin edible coating on postharvest quality and storability of red bell peppers. Postharvest Biology and Technology 96, 106—109.

Rebouillat, S., Ortega-Requena, S., 2015. Potential applications of milk fractions and valorization of dairy by-products: a review of the state-of-the-art available data, outlining the innovation potential from a bigger data standpoint. Journal of Biomaterials and Nanobiotechnology 6 (03), 176–203.

Roselló-Soto, E., Koubaa, M., Moubarik, A., Lopes, R.P., Saraiva, J.A., Boussetta, N., Grimi, N., Barba, F.J., 2015. Emerging opportunities for the effective valorization of wastes and by-products generated during olive oil production process: non-conventional methods for the recovery of high-added value compounds. Trends in Food Science & Technology 45 (2), 296–310.

Saberi, B., Golding, J.B., Marquez, J.R., Pristijono, P., Chockchaisawasdee, S., Scarlett, C.J., Stathopoulos, C.E., 2018. Application of biocomposite edible coatings based on pea starch and guar gum on quality, storability and shelf life of "Valencia oranges". Postharvest Biology and Technology 137, 9–20.

Salaberria, A.M., Labidi, J., Fernandes, S.C.M., 2014. Chitin nanocrystals and nanofibers as nano-sized fillers into thermoplastic starch-based biocomposites processed by melt-mixing. Chemical Engineering Journal 256, 356–364.

Sánchez-González, L., Pastor, C., Vargas, M., Chiralt, A., González-Martínez, C., Cháfer, M., 2011. Effect of hydroxypropyl methylcellulose and chitosan coatings with and without bergamot essential oil on quality and safety of cold-stored grapes. Postharvest Biology and Technology 60, 57–63.

Sanchez-Olivares, G., Sanchez-Solis, A., Calderas, F., Alongi, J., 2017. Keratin fibres derived from tannery industry wastes for flame retarded PLA composites. Polymer Degradation and Stability 140, 42–54.

Sánchez-Safont, E.L., Aldureid, A., Lagarón, J.M., Gámez-Pérez, J., Cabedo, L., 2018. Biocomposites of different lignocellulosic wastes for sustainable food packaging applications. Composites Part B 145, 215–225.

Shankar, S., Reddy, J.P., Rhim, J.W., Kim, H.Y., 2015. Preparation, characterization, and antimicrobial activity of chitin nanofibrils reinforced carrageenan nanocomposite films. Carbohydrate Polymers 117, 468–475.

Sila, A., Bougatef, A., 2016. Antioxidant peptides from marine by-products: isolation, identification and application in food systems. A review. Journal of Functional Foods 21, 10–26.

Siripatrawan, U., Noipha, S., 2012. Active film from chitosan incorporating green tea extract for shelf life extension of pork sausages. Food Hydrocolloids 27, 102–108.

Smuda, S.S., Mohsen, S.M., Olsen, K., Aly, M.H., 2018. Bioactive compounds and antioxidant activities of some cereal milling by-products. Journal of Food Science & Technology 55 (3), 1134–1142.

Sogut, E., Seydim, A.C., 2018. The effects of chitosan and grape seed extract-based edible films on the quality of vacuum packaged chicken breast fillets. Food Packaging and Shelf Life 18, 13–20.

Song, N.B., Lee, J.H., Al Mijan, M., Song, K.B., 2014. Development of a chicken feather protein film containing clove oil and its application in smoked salmon packaging. LWT-Food Science and Technology 57, 453–460.

Sung, S.H., Chang, Y., Han, J., 2017. Development of polylactic acid nanocomposite films reinforced with cellulose nanocrystals derived from coffee silverskin. Carbohydrate Polymers 169, 495–503.

Tatullo, M., Simone, G.M., Tarullo, F., Irlandese, G., Vito, D.D., Marrelli, M., Santacroce, L., Cocco, T., Ballini, A., Scacco, S., 2016. Antioxidant and antitumor activity of a bioactive polyphenolic fraction isolated from the brewing process. Scientific Reports 6, 36042.

Thakur, R., Saberi, B., Pristijono, P., Golding, J., Stathopoulos, C., Scarlett, C., Bowyer, & M., Vuong, Q., 2016. Characterization of rice starch-ι-carrageenan biodegradable edible film. Effect of stearic acid on the film properties. International Journal of Biological Macromolecules 93, 952–960.

Teixeira, A., Baenas, N., Dominguez-Perles, R., Barros, A., Rosa, E., Moreno, D., Garcia-Viguera, C., 2014. Natural bioactive compounds from winery by-products as health promoters: a review. International Journal of Molecular Sciences 15 (9), 15638–15678.

Toldrá, F., Mora, L., Reig, M., 2016. New insights into meat by-product utilization. Meat Science 120, 54–59.

Tulamandi, S., Rangarajan, V., Rizvi, S.H., Singhal, R.S., Chattopadhyay, S.K., Saha, N.C., 2016. A biodegradable and edible packaging film based on papaya puree, gelatin, and defatted soy protein. Food Packaging and Shelf Life 10, 60–71.

Turina, A.V., Nolan, M.V., Zygadlo, J.A., Perillo, M.A., 2006. Natural terpenes: self-assembly and membrane partitioning. Biophysical Chemistry 122, 101–113.

Tyug, T.S., Prasad, K.N., Ismail, A., 2010. Antioxidant capacity, phenolics and isoflavones in soybean by-products. Food Chemistry 123 (3), 583–589.

Vargas, M., Albors, A., Chiralt, A., González-Martínez, C., 2006. Quality of cold-stored strawberries as affected by chitosan-oleic acid edible coatings. Postharvest Biology and Technology 41 (2), 164–171.

Vargas, M., Albors, A., Chiralt, A., 2011. Application of chitosan-sunflower oil edible films to pork meat hamburgers. Procedia Food Science 1, 39–43.

Vázquez, G., Fontenla, E., Santos, J., Freire, M.S., González-Álvarez, J., Antorrena, G., 2008. Antioxidant activity and phenolic content of chestnut (*Castanea sativa*) shell and eucalyptus (*Eucalyptus globulus*) bark extracts. Industrial Crops and Products 28 (3), 279–285.

Vijayalaxmi, S., Jayalakshmi, S.K., Sreeramulu, K., 2015. Polyphenols from different agricultural residues: extraction, identification and their antioxidant properties. Journal of Food Science & Technology 52 (5), 2761–2769.

Wang, N., Gao, Y.Z., Wang, P., Yang, S., Xie, T.M., Xiao, Z.G., 2016. Effect of microwave modification on mechanical properties and structural characteristics of soy protein isolate and zein blended film. Journal of Food Science 34, 180–188.

Yu, Z., Sun, L., Wang, E., Zeng, W., Mustapha, A., Lin, M., 2018. Soy protein-based films incorporated with cellulose nanocrystals and pine needle extract for active packaging. Industrial Crops and Products 112, 412–419.

Yuan, B., Lu, M., Eskridge, K.M., Hanna, M.A., 2018. Valorization of hazelnut shells into natural antioxidants by ultrasound-assisted extraction: process optimization and phenolic composition identification. Journal of Food Process Engineering 41, e12692.

Zhao, Y., Zhang, Y., Lindström, M.E., Li, J., 2015. Tunicate cellulose nanocrystals: preparation, neat films and nanocomposite films with glucomannans. Carbohydrate Polymers 117, 286–296.

PART

Innovative food technologies and nutrients bioavailability

III

CHAPTER

Development of functional foods by traditional food processes

7

Joanna Harasym, DSc, PhD, MSc [1,2], Urszula Kaim[1,2], Ludmiła Bogacz-Radomska[1,2], Remigiusz Oledzki[1,2]

[1]*Adaptive Food Systems Accelerator, Wrocław University of Economics, Wrocław, Poland;*
[2]*Department of Biotechnology and Food Analysis, Wrocław University of Economics Wrocław, Poland*

1. Introduction

Traditional food processing methods were originally created by the natural human need for nutritional food delivering. Modified or discarded by the modern food industry, they lost their primary impact as processing technologies for human food manufacturing but still consist of the unexplored source of methods and procedures capable to create not only nutritious but also functional food products.

In a changing world, the concept of food was also changing a number of times. Not so long ago, food in Western civilization was treated only as a factor necessary to maintain life being, satisfy hunger, or ensure the development of the body. However, with the progress of civilization, the expansion of knowledge, the industrial revolution, and the emergence of new processing and preservation techniques, the role of food in Western societies has also acquired a completely new meaning. Awareness of this role deepened in the course of biological, civilizational, and cultural development of man.

This consciousness developed separately in each of the man-made civilizations because they were spatially separated. This separation has also led to the fact that Western civilization is now "discovering" knowledge about the impact of food anew, despite the enormous achievements of the East, Africa, or even Medieval Europe. Additionally, this perception is now polarized by the prism of the primacy of the economy. At present, the global functional food market is valued at EUR 34 billion, which means that for a long time to come, in Western civilization, the prospect of earning money will shape consumer habits (Leatherhead food research, 2016). It is widely recognized that the consumption of functional food is a remedy for the overwhelming number of cases of diseases from the group of diseases of civilization—those that are directly related to lifestyles.

The food production is one of the most important industries. The development of methods of obtaining food while at the same time ensuring its safety and efficiency

Sustainability of the Food System. https://doi.org/10.1016/B978-0-12-818293-2.00007-0

of the technological process is the greatest challenge faced by both food producers and scientists related to food technology (Pouliot et al., 2014).

Currently, the development of food and nutrition sciences is a source of information on the mechanisms of the influence of food components on human metabolism. This knowledge is also the reason for the growth of the special food market, in respect of which consumer expectations are built on the basis of scientifically proven activities.

Scientific research focuses on reducing the risk of incidence of the so-called civilization diseases, chronic and cardiovascular diseases, obesity, cancer, osteoporosis, and non−insulin-dependent diabetes mellitus (Mertz, 2008).

Numerous scientific studies have shown that bioactive ingredients are helpful in preventing many diseases. Based on this knowledge, pharmaceutical products such as herbal medicines and food products are developed. There is an increase in the variety of food products to meet the needs of different target groups, e.g., the elderly, pregnant and breastfeeding women, infants, young children, and athletes. In an ageing population in Europe and North America, it is also important for many elderly people that the products they consume can provide a balanced diet without changing their long-term habits (Harasym et al., 2016).

A large number of compounds contained in food affect the occurrence of coronary heart disease, which is discussed in the previous paragraph. The effects of most of them have been known for a long time, such as saturated fats or salt, and others have been identified over the past few decades. At the same time, a wide range of active compounds are an inexhaustible source of the creative approach to food production.

The development of processing methods forced by demographic changes and the progress of civilization expressed, among others, in the form of the industrial revolution resulted in the creation of a number of food products with different levels of processing from unprocessed to highly processed food.

The industrialization of agriculture and food production increased significantly after World War II, being the form to address the global hunger and secure the efficient and safe food supply. Unfortunately, the global shift toward such model of food systems in the past 70 years has resulted in many environmental costs. Industrialized food production is highly mechanized and concentrated, based on harsh chemical utilization such as pesticides, fertilizers, and nontherapeutic antibiotics. Even though, in the countries or regions that mostly benefit from industrialized food production, the overall safety is monitored, there are still sickens coming from contaminated food annually, whereas the antibiotics that now are routinely fed to livestock start to promote the mutation, resulting in antibiotic-resistant bacteria appearance. The agricultural pesticide leakages created the death zones in both open water and groundwater adjacent to crop fields, resulting in drinkable water contamination. Therefore, the implementation of sustainable food production is of high importance.

Currently, the food production in Western civilization is shifting toward pro-health impact of food. The trend initiated by increasing awareness of the properties of food and the impact of food on human health is gaining momentum, becoming an important market position. In the mid-1980s, in Japan and then in the United States, products known as functional food began to appear on the market (Kudełka, 2011), which is not particularly astonishing given the approach to food in Asia and the level of development of food production, and Japan remains the market leader.

The possibility of recognizing a product as a functional food requires scientific research. Its purpose is to show the existence of a link between the consumption of food and a specific prohealth effect. Preliminary research concerns the knowledge of the mechanisms of functioning of active ingredients contained in food, whereas proper nutritional research is carried out on animals and in vitro models and in the final stage on humans to verify the assumptions of functionality.

The functional food market sustainable development targets to protect the safety of the people consuming food that should be healthy. If one of the factors of sustainable food supply will be the contribution to the public health, the investment in the functional food products development seems to be reasonable practice to increase the overall sustainability of food systems. The benefits coming from functional food production and consumption can be compared with different approaches promoting healthy living, indicating the huge contribution of proper diet, especially dedicated food products (functional foods). However, the evidence coming from different studies (Wichienchot and Ishak, 2017) indicated the more global aspect, which should be also addressed when talking about health food systems. The sustainability is intrinsically linked to policies, cultural diversity, and sometimes traditional approaches to foods and food practices. To increase the sustainable approach in functional food production is to balance the choice between development of innovative functional products and environmentally friendly processing. Both biodiversity of regions and cultural heritage need to be taken into account, as well as citizen/consumers rights versus development needs of food industry when planning the sustainable development of this food sector development.

Multicenter studies showed a significant advantage of plant food grown using organic methods over products from typical cultivation (http://research.ncl.ac.uk/nefg/QOF/page.php?page=1). Conventional and organic animal husbandry also shows a clear variation in the nutritional value of the food products obtained in this way. An international team of experts led by scientists from the University of Newcastle, United Kingdom, has shown that both organically sourced milk and meat contain about 50% more beneficial omega-3 fatty acids than conventionally produced products. Significant differences were found between milk and meat obtained by organic and conventional farming methods, especially in terms of the content of fatty acids, as well as some minerals and antioxidants (Średnicka-Tober et al., 2016). Van Asselt et al. (2015) studied the milk production in Holland and concluded that providing raw organic milk is far more sustainable than pasteurized one (Van Asselt et al., 2015).

The traditional food preparation processes and their impact on nutrients level in food products still consist of unexplored territory delivering inspiring knowledge, which can result in novel and nutritious functional products. Enzymatic conversion using internal or external sources of enzymes had been traditionally applied for both food conservation and its nutritional level enhancement. Specific thermal treatments allowed to obtain easily digestible food with high level of microelements, vitamins, and antioxidants. Chemical processing using natural acids and alkali facilitates hydrolysis of biopolymers liberating highly bioactive compounds from internal plant matrix.

The objective of this chapter is to provide the deep insight into the variety of just existing traditional methods of food processing, which can supply the functional food products without necessity of high and advanced processing. Those methods offer the sustainable approach for functional food manufacturing with nutritious ingredient preservation or increase along with biodiversity protection. The traditional processes of nutritious grain product preparation (germination, malting), exploitation of natural microflora of raw materials (sourdough, fermentation), natural chemical compounds usage (spicing, pickling), or physical preparation (stone milling, roasting) deliver functional food with minimal industrial processing efforts, which makes that methods very sustainable alternative.

2. Grain product preparation

2.1 Germination of seeds

Recent studies show that germination can enhance both the nutritional and medicinal values of seeds, legumes, and cereals and decrease the amount of antinutritional and indigestible factors (Gan et al., 2017). During germination, the activity of endogenous phytase increases especially in cases of such cereals such as maize and sorghum, with a lower endogenous phytase activity than rye, wheat, triticale, buckwheat, and barley (Egli et al., 2002; Nkhata et al., 2018). In a result of germination, also tannins and other polyphenols in legumes may be reduced by formation of polyphenol complexes with proteins and the gradual degradation of oligosaccharides, which may provide to facilitate iron absorption (Camacho et al., 1992). The activity of α-amylase is increased during germination, which is responsible for reducing the viscosity of cereal and legumes suspensions by hydrolysis of amylose and amylopectin to dextrins and maltose (Gibson et al., 1998).

Sprouting can significantly increase the amount of polyphenol in seeds, cereals, and pseudocereals. Alvarez-Jubete et al. (2010) reported that polyphenols content in quinoa, buckwheat, and wheat was doubled and in case of amaranth even quadrupled (Alvarez-Jubete et al., 2010). Besides, germination improves digestion of starch and proteins, increases content of vitamins B, C, and E, and decreases the amount of protease inhibitors and lecithin (Gan et al., 2017; Chavan, and Kadam, 1989). Thus, germinated seeds, rich in bioactive compounds, may be consumed as functional foods to prevent and treat chronic diseases (Chu et al., 2019).

Germination can be an inexpensive and effective technology for improving the quality of food, by enriching its nutritional value (Chavarín-Martínez et al., 2019). The applications of germinated ingredient in food preparation are very diverse, such as the use of germinated legumes and cereals for flour, beverage, and weaning food manufacturing (Rusydi et al., 2011). Gawlik-Dzik et al. (2014) used germinated, dried, and powdered broccoli sprouts as an addition to wheat bread formulation. The bread obtained was rich in phenolic compounds and enzyme effectors and provided a higher level of free amino groups and resistant starch (Gawlik-Dziki et al., 2014). Xu et al. (2014) used germinated tartary buckwheat as an additive to steamed wheat bread. The bread has less starch and is enriched in Fe, Mn, and several of bioactive compounds such as protocatechuic acid and quercetin. They also reported a significant increase in antioxidant activity in bread (Xu et al., 2014).

Cáceres et al. (2019) used germinated brown rice for the developed multifunctional yogurt-like product. They achieve the improvement of bioactive compounds content such as phenolic compounds and antioxidants. Such product is suggested as a replacement of traditional yogurt for vegans or people interested in reducing animal protein intake (Cáceres et al., 2019).

2.2 Malting process

Malting is a controlled drying process of germinated grains. Drying, mostly using hot air, allows to stops the enzymatic processes in the germinated grain and induces/adds phytochemical alterations and the typical flavor characteristics of malt. Malt is considered as a whole grain, which consumption is commonly related with health benefits (Nelson et al., 2013; Gupta et al., 2010). Malting promotes the development of hydrolytic enzymes, which are not present in the nongerminated, raw grain (Gawande and Babar, 2018). Malting can improve the antioxidant activity, total phenolic content, phenolic compounds, and carotenoid content of grains. It also provides to increase in the bioactive components and fatty acid composition, such as catechin, caffeic acid, and quercetin content and linoleic acid (Özcan et al., 2018).

Malting also changes the mineral and physiochemical characteristic of grains. The mineral content of wheat malt and barley malt shows a significant increase in content of calcium, iron, potassium, zinc, and magnesium and in water-soluble fiber, oligosaccharides, and resistant starch. Thus, malts can be successfully used in functional foods as prebiotic additives (Nelson et al., 2013; Hingade et al., 2019).

Malt is used all over the world for manufacturing beverages, malt-enriched food products, and industrial purposes in medicine and manufacturing baby food (Alam et al., 2007; Carvalho et al., 2016). A number of studies have indicated that malting is a process that transforms the grain into a more palatable and nutritionally richer form (Donko et al., 2012).

Molinari and others (2018) used tartary buckwheat malt for enhancing polyphenols and flavonoids content and antioxidant capacity of gluten-free cookies. The 30% amount of tartary buckwheat malt resulted in significant increase of total phenolic, quercetin, rutin, and antioxidant activity. They reported also lower caloric value and higher content of protein, fiber, and resistant starch, compared with control (Molinari et al., 2018).

Malts can be used also in weaning food formulations manufacturing. Kulkarni and others (1991) used the malted sorghum and green gram in weaning food formulations obtaining improved viscosity and nutritive value, especially the protein, available lysine, and vitamin C content (Kulkarni et al., 1991).

3. Exploitation of natural microflora of raw materials

3.1 Sourdough

Sourdough fermentation, recognized as a reliable tool to improve nutritional, functional, technological, and sensory properties of wheat and nonwheat flours, is related with metabolic activities of two groups of microorganisms: yeasts and lactic acid bacteria (LAB). It has been shown that enzymatic activities of LAB are responsible for an increase in functional compounds content of phenolic compounds and soluble fibers, recommended to prevent several diseases and regulate energy intake and satiety, lowering total and low-density lipoprotein (LDL) cholesterol levels and helping with glycemic control (Kulkarni et al., 1991; Montemurro et al., 2019; Curiel et al., 2015; Chater et al., 2015). Though starch digestibility and formation of organic acids are reduced, sourdough fermentation lowers the glycemic index of bread. The sourdough wholemeal wheat bread in comparison with white wheat bread, whole meal wheat bread, and whole meal bread made with xylanase shows the lowest postprandial glucose and insulin response (D'Alessandro and Pergola, 2014).

Phytate, commonly presented in whole grains, legumes, oilseeds, and nuts, is strongly bound to metal cations such as potassium, iron, calcium, magnesium, manganese, and zinc, which make them insoluble and unavailable as nutritional factors (Schlemmer et al., 2009; Bohn et al., 2008). Sourdough fermentation is an efficient process for reducing phytate content in whole wheat bread. Reduction of the pH value, caused by microbiological activity, provided the beneficial conditions for the endogenous cereal phytase activity (Lopez et al., 2001; Reale et al., 2007).

Consumer demand for functional food of improved nutritional values, which provides health benefits, or corresponding to "clean label" trend is evolving (Rizzello et al., 2017; Brandt, 2007). It has been shown that sourdough fermentation, of alternative flours, from pseudocereals or legumes, can improve baking, sensory, and nutritional values (Bartkiene et al., 2011)). In products, where legume flours were added to wheat bread formulation, the use of sourdough fermentation leaded to increase of free amino acid and γ-aminobutyric acid (GABA) concentration (Coda et al., 2010; Rizzello et al., 2014), and antioxidant and phytase activities (Peñaloza-Espinosa et al., 2011) and improved nutritional quality by increase in protein digestibility (Peñaloza-Espinosa et al., 2011; Bartkiene et al., 2011).

3.2 Fermentation

Fermentation enables food preservation and improvement based on biological processes. This technique applied microorganisms that transform substrates into valuable products to satisfy the consumer needs in a field of food, energy, and chemical industry.

Fermented food and beverages exhibit enhanced shelf life and biological value resulting from the presence of bioactive compounds, vitamins, and microorganisms (Rezac et al., 2018). These compounds benefit human health due to antioxidant activity and improvement of digestion, reducing the risk of diabetes and cardioprotective properties (Archer and Halami, 2017; Sivamaruth et al., 2018; de Carvalho et al., 2018). Fermented food shows the antipathogenic, antibiofilm, and technological properties (Khusro et al., 2019).

Fermentation process is mainly induced by LAB combined with acetic acid bacteria and yeasts (García et al., 2019). LAB were used at biotransformation of glucosinolates, phenolics, and sugars in retorted broccoli pure (Ye et al., 2019). In Asia and West Africa, *Bacillus* species are applied in the fermentation of soybeans and locust beans (Kimura and Yokoyama, 2019; Devanthi and Gkatzionis, 2019). Legume sprouts were fermented by LAB toward enhancing the microbiological safety and content of isoflavonoids (Budryn et al., 2019). Regarding the substrate, there are nine groups of fermented food: miscellaneous fermented products, alcoholic beverages, meat products, fish products, milk, roots/tubers, legumes, vegetables, and fermented cereals (Tamang et al., 2016).

Nowadays that process is applied to formulate functional and novel food. Dry-fermented sausage with incorporated dietary fiber exhibited higher antioxidant capacity (Pérez-Burillo et al., 2019). Quinoa and buckwheat cooked seeds were fermented by LAB toward phenolic acids and tyrosols (Rocchetti et al., 2019). Traditional fermented food like fermented barley bran in South Korea was enriched with β-glucan, resulting in protection of gastric mucosal injury (Jeong et al., 2019). During sourdough fermentation by lactic acid bacteria were obtained levans (Shi et al., 2019). Cofermentation of sorghum and pigeon pea was used to develop a weaning food (Adebayo-Oyetoro et al., 2019). The process of fermentation was applied to degrade of allergen in soybean meal and improve its protein quality (Li et al., 2019).

4. Natural chemical compounds usage

4.1 Spicing

Spices are natural vegetable products, used to improve the taste, aroma, and color of the dishes. Spices exhibit antioxidant or bacteriostatic, bactericidal, antifungal, and anticancer properties; therefore, they are applied at the treatment of many diseases. In addition, they affect positive the digestive, nervous, reproductive, and blood systems. Spices extend the shelf life of products by inhibition of microorganisms and oxidation processes. They contain many valuable minerals and vitamins, such as potassium, calcium, phosphorus, magnesium, iron, β-carotene, and vitamin C (Akolade et al., 2019; Khazdair et al., 2019).

Among them, those support digestion system are Ceylon cinnamon (*Cinnamomum verum*), cloves (*Caryophylli flos*), coriander (*Coriandrum sativum*), nutmeg (*Myristica fragrans*), and cardinal malabar (*Elettaria cardamomum*). In the treatment of digestive tract disorders, spices such as caraway seeds (*Carum carvi*) or garden dill (*Anethum graveolens*) are used (Kazemipoor and Cordell, 2015; Fazel et al., 2017).

Healing properties are also exhibited by oregano (*Origanum vulgare*), bay leaf (*Laurus nobilis*), basil (*Ocimum basilicum*), thyme (*Thymus vulgaris*), and pepper (*Piper*). These spices have antiinflammatory properties and are therefore recommended in rheumatoid arthritis (Balusamy et al., 2018; Sharma et al., 2012; Seibel et al., 2018; Jungbauer et al., 2012). Medicinal ginger (*Zingiber officinale*) has an immune system strengthening function. Properties of fresh or dried ginger are applied to avoid nausea in pregnant women, at the treatment of the motion sickness, bronchitis, hypertension, runny nose, asthma, dental and gingivitis, constipation, and nervous system diseases (Srinivasan, 2017). Due to sulfur presence, garlic (*Allium sativum*) exhibits the strengthening properties. Garlic has antioxidant, antibacterial, antifungal, antiatherosclerotic, blood pressure, anticoagulant, and antineoplastic properties (Suleria et al., 2015).

Turmeric (*Curcuma*) has antiinflammatory, antioxidant, anticancer, and antidiabetic effects. It is applied to the treatment of digestive system diseases, especially hepatic disorders. Turmeric lowers total blood cholesterol and its LDL fractions and stops the replication of HIV-1 (Salehi et al., 2019; Shakeri et al., 2019; Meng et al., 2018).

Spices are widely added to different dishes, according to recipes and personal likeness, consisting easy source of bioactive functional compounds.

4.2 Pickling/marinating

Marinating is one of the oldest and most effective methods of food processing and preparation for long-term storage. The most commonly used chemical substances in the marinating process are acetic, citric, malic, and tartaric acids. Marinating makes it possible to fix food by increasing its acidity (lowering the pH), which inhibits the development of microorganisms (Chung et al., 2017). The health benefits of marinated foods are linked to the abundance of vitamins, minerals, and antioxidants in such products. Due to this fact, the consumption of marinated food is attributed to a number of therapeutic activities related to the control of diabetes, improvement of digestion processes, or liver function support (Lawson and Hunsaker, 2018). Since fresh fruit and vegetables are used in the marinating process, both the amount and activity of antioxidants present in these raw materials can be fully maintained. Plant products are thus marinated as a rich source of antioxidants. It has been shown that, e.g., the fruit marinades of the *Decalepis hamiltonii* plant of the gentian order show particular antioxidant properties and high polyphenol content (Mallikarjuna and Dharmesh, 2018). It is indicated that marinades can stimulate the development of beneficial bacteria in the gastrointestinal tract, thus supporting food digestion processes. The consumption of marinated fruits and vegetables is particularly important in the situation of antibiotic therapy, which destroys invasive bacteria and kills human-friendly bacteria (Dallal et al., 2017). A special role in stimulating digestive processes is attributed to marinades of Indian gooseberry or amla fruit (*Phyllanthus emblica*). For this reason, it is recommended that these products are eaten as a starter before the severely digestible or before the last daily meal.

5. Physical preparation

5.1 Roasting

Roasting is applied at cooking and food processing and is the preservation method of food. The roasting process results in aromas and flavors creation. Because of this, roasting is applied especially at coffee, cocoa, and tea production so that the key aroma compounds are obtained (Giacalone et al., 2019; Rocha et al., 2017; Morikawa et al., 2019). Roasting bases on heating raw materials at a high temperature, below their melting point, to cause physical or chemical changes. Roasting coffee is a complex process. The key reactions applied in aroma formations are Maillard reaction, carbohydrate caramelization, and oxidation and Strecker degradation (Wongsa et al., 2019). Roasting may cause defects resulting in lack of aroma of the coffee brew because the coffee aroma is related to the time and temperature during roasting of coffee beans (Baggenstoss et al., 2008; Fisk et al., 2012; Masi et al., 2013; Yang et al., 2016). There are several prohealthy properties of roasted coffee consumption. Due to chlorogenic acids, coffee is a potential functional drink that may prevent Alzheimer's disease and exhibit antioxidant properties (Kondeva-Burdina et al., 2019). The plasma-oxidized LDL and α-tocopherol were improved by the consumption of dark-roast coffee (Hochkogler et al., 2019).

Cocoa beans are processed mainly using the roasting process, which affects the color and texture of beans, flavor compounds, acidity, and lipid content. The key factors of roasting conditions are the time of exposure and temperature (Yang et al., 2016).

Roasted green tea is a common drink in Japan, and it has characteristic flavor resulting from Maillard reactions and caramelization. Compared with the nonroasted tea, it has a milder taste and lower caffeine content in tea (Rocha et al., 2017).

Roasting the rapeseed improved the flavor of rapeseed oil and increased the antioxidant content (Gracka et al., 2016). Also the flavor profile of dark chocolate depends on roasting temperature (Hinneh et al., 2019). Roasting has a negative influence on protein and vitamin C contents of sweet potato and simultaneously increased the antioxidant potential (Hou et al., 2018). The flour of roasted Kalonji seeds exhibited higher water and oil absorption. However, roasting compounded the viscosity, emulsion, and foaming characteristics of that flour (Jan et al., 2019). Different nuts such as hazelnuts, macadamia nuts, walnuts, almonds, and pistachios are commercially roasted. However, roasting temperatures only from low to middle (120−160°C) affect positive the sensory properties of nuts (Schlörmann et al., 2015). Pistachios exhibit antitumor properties and contain potassium (that lower blood pressure), lutein, and zeaxanthin (that reduce the risk of macular degeneration). Therefore, the increase in their consumption via aroma enhancement is an important challenge.

5.2 Stone milling method

Stone milling is a single-stream milling method so that the grain is crushed between two millstones and stay in one batch to the end of the milling process (Miller Jones et al., 2015). The milling stones consist of igneous rocks such as rhyolitic ignimbrite and basalt. Stone milling generates temperature up to 90°C between the stones; however, that depends on rotary speed. That may affect the protein, starch, and unsaturated fatty acids (Guerrini et al., 2019). Stone milling influenced the chemical composition and antioxidant potential of chosen flour so that the number of phenolic compounds and antioxidant acidity increase significantly compared with other milling methods. Stone milling caused high share of proteins, minerals, and fibers in flour composition (Liu et al., 2018). Stone milling affects the bioavailability of calcium, iron, and zinc, resulting from large particle size and high extraction level (Xue et al., 2016; Hurrel et al., 2010).

6. Conclusions

Traditional processes of food processing have been invented to provide the nutritional food and moreover verified over many decades of usage. The traditional knowledge is now appreciated and is tried to exploit on a wider scale. Not all traditional methods can be scaled up due to time and resource consuming routine. However, it should not be necessary, as small scale could be successfully applied in craft manufacturers providing food on the local scale. Such form has several benefits as it supports local economies, utilizes local resources, and minimizes the storage time and, by this, the preservatives and additive usage. The food manufactured from fresh local raw materials can provide nutritional benefits resulting in well-balanced nutrition. Traditional processing methods applied to new raw materials can otherwise deliver novel food products rich in nutrients, providing health benefits beyond the typical ingredients. Such an approach can result in innovative functional food products created in a sustainable way.

Acknowledgments

The study was supported by National Science Centre, Poland, through grant no. 2018/02/X/NZ9/01693.

Authors thanks for support of CREABIOECON Flagship InterTeam project (JH) and QCON-Gfree Young Researcher project (UK) of INTEREKON grant financed by the Ministry of Science and Higher Education in Poland under the program "Regional Initiative of Excellence" 2019–2022 project number 015/RID/2018/19 total funding amount 10 721 040,00 PLN.

References

Adebayo-Oyetoro, A.,O., Adeyeye, S.A.,O., Olatidoye, O.,P., Ogundipe, O.,O., Adenekan, O.,E., 2019. Effect of co-fermentation on the quality attributes of weaning food produced from sorghum (*Sorghum bicolor*) and pigeon pea (*Cajanus cajan*). Journal of Culinary Science & Technology 17, 118–135.

Akolade, J.,O., Na'Allah, A., Sulyman, A.,O., Abdulazeez, A.,T., Atoti, A.,O., Isiaku, M.,B., 2019. Antidiabetic screening of phenolic-rich extracts of selected medicinal spices. Iranian Journal of Science and Technology Transaction A-Science 43, 357–367.

Alam, M.Z., Haider, S.A., Paul, N.K., 2007. Yield and yield components of barley (*Hordeum vulgare* L.) cultivars in relation to nitrogen fertilizer. International Journal of Applied Sciences 3, 1022–1026.

Alvarez-Jubete, L., Wijngaard, H., Arendt, E.K., Gallagher, E., 2010. Polyphenol composition and in vitro antioxidant activity of amaranth, quinoa buckwheat and wheat as affected by sprouting and baking. Food Chemistry 119, 770–778.

Archer, A.,C., Halami, P.,M., 2017. Fermented foods, microbiota and human health. In: Kalia, V., Shouche, Y., Purohit, H., Rahi, P. (Eds.), Mining of Microbiota Wealth and Meta Genomics. Springer, Singapore, pp. 301–331.

Baggenstoss, J., Poisson, L., Kaegi, R., Perren, R., Escher, F., 2008. Coffee roasting and aroma formation: application of different time-temperature conditions. Journal of Agricultural and Food Chemistry 56, 5836–5846.

Balusamy, S.,R., Perumalsamy, H., Huq, A., Balasubramanian, B., 2018. Anti-proliferative activity of *Origanum vulgare* inhibited lipogenesis and induced mitochondrial mediated apoptosis in human stomach cancer cell lines. Biomedicine & Pharmacotherapy 108, 1835–1844.

Bartkiene, E., Joudeikiene, G., Vidmantiene, D., Viskelis, P., Urbonaviciene, D., 2011. Nutritional and quality aspects of wheat sourdough bread using *L. luteus* and *L. angustifoliusflours* fermented by *Pediococcus acidilactici*. International Journal of Food Science and Technology 46, 1724–1733.

Bohn, L., Meyer, A.S., Rasmussen, S.K., 2008. Phytate: impact on environment and human nutrition. A challenge for molecular breeding. Journal of Zhejiang University – Science B 9, 165–191.

Brandt, M.,J., 2007. Sourdough products for convenient use in baking. Food Microbiology 24, 161–164.

Budryn, G., Klewicka, E., Grzelczyk, J., Gałązka-Czarnecka, I., Mostowski, R., 2019. Lactic acid fermentation of legume seed sprouts as a method of increasing the content of isoflavones and reducing microbial contamination. Food Chemistry 285, 478–484.

Cáceres, P.,J., Peñas, E., Martínez-Villaluenga, C., García-Mora, P., Frías, J., 2019. Development of a multifunctional yogurt-like product from germinated brown rice. LWT – Food Science and Technology 99, 306–312.

Camacho, L., Sierra, C., Campos, R., Guzman, E., Marcus, D., 1992. Nutritional changes caused by germination of legumes commonly eaten in Chile. Archivos Latinoamericanos de Nutricion 42, 283–290. PMID:1342162.

Carvalho, D.O., Goncalves, L.M., Guido, L.F., 2016. Overall antioxidant properties of malt and how they are influenced by the individual constituents of barley and the malting process. Comprehensive Reviews in Food Science and Food Safety 15, 927–943.

Chater, P.,I., Wilcox, M.,D., Pearson, J.,P., Brownlee, I.,A., 2015. The impact of dietary fibres on the physiological processes governing small intestinal digestive processes. Bioactive Carbohydrates and Dietary Fibre 6, 117–132.

Chavan, J.,K., Kadam, S.,S., 1989. Nutritional improvement of cereals by sprouting. Critical Reviews in Food Science and Nutrition 28, 401–437.

Chavarín-Martínez, C.,D., Gutiérrez-Dorado, R., Perales-Sánchez, J.X.K., Cuevas-Rodríguez, E.O., Milán-Carrillo, J., Reyes-Moreno, C., 2019. Germination in optimal conditions as effective strategy to improve nutritional and nutraceutical value of underutilized Mexican blue maize seeds. Plant Foods for Human Nutrition 74, 192–199.

Chu, C., Yan, N., Du, Y., Liu, X., Chu, M., Shi, J., Zhang, H., Liu, Y., Zhang, Z., 2019. iTRAQ based proteomic analysis reveals the accumulation of bioactive compounds in Chinese wild rice (*Zizania latifolia*) during germination. Food Chemistry. https://doi.org/10.1016/j.foodchem.(2019).03.092 (in press): Accepted Manuscript.

Chung, I.-M., Oh, J.-Y., Kim, S.-H., 2017. Comparative study of phenolic compounds, vitamin E, and fatty acids compositional profiles in black seed-coated soybeans (*Glycine Max* (L.) Merrill) depending on pickling period in brewed vinegar. Chemistry Central Journal 11, 64.

Coda, R., Rizzello, C.G., Gobbetti, M., 2010. Use of sourdough fermentation and pseudo-cereals and leguminous flours for the making of a functional bread enriched of γ-amino-butyric acid (GABA). International Journal of Food Microbiology 137, 236–245.

Curiel, J.,A., Coda, R., Centomani, I., Summo, C., Gobbetti, M., Rizzello, C.G., 2015. Exploitation of the nutritional and functional characteristics of traditional Italian legumes: the potential of sourdough fermentation. International Journal of Food Microbiology 196, 51–61.

Dallal, M.M.,S., Zamaniahari, S., Davoodabadi, A., Hosseini, M., Rajabi, Z., 2017. Identification and characterization of probiotic lactic acid bacteria isolated from traditional Persian pickled vegetables. GMS Hygen and Infection Control. https://doi.org/10.3205/dgkh000300.

de Carvalho, N.,M., Costa, E., Silva, S., Pimentel, L., Fernandes, T.,H., Pintado Esteves, M., 2018. Fermented foods and beverages in human diet and their influence on gut microbiota and health. Fermentatio 4, 90.

Devanthi, P.V.P., Gkatzionis, K., 2019. Soy sauce fermentation: microorganisms, aroma formation, and process modification. Food Research International 120, 364–374.

Donkor, O.,N., Stojanovska, L., Ginn, P., Ashton, J., Vasiljevic, T., 2012. Germinated grains-sources of bioactive compounds. Food Chemistry 135, 950–959.

D'Alessandro, A., De Pergola, G., 2014. Mediterranean diet pyramid: a proposal for Italian people. Nutrients 6, 4302–4316.

Egli, I., Davidsson, L., Juillerat, M.-A., Barclay, D., Hurrell, R., 2002. The influence of soaking and germination on the phytase activity and phytic acid content of grains and seeds potentially useful for complementary feeding. Journal of Food Science 67, 3484–3492.

Fazel, N., Pejhan, A., Taghizadeh, M., Tabarraei, Y., Sharifi, N., 2017. Effects of *Anethum graveolens* L. (Dill) essential oil on the intensity of retained intestinal gas, flatulence and pain after cesarean section: a randomized, double-blind placebo-controlled trial. Journal of Herbal Medicine 8, 8–13.

Fisk, I.,D., Kettle, A., Hofmeister, S., Virdie, A., Kenny, J.,S., 2012. Discrimination of roast and ground coffee aroma. Flavour 1, 14.

Gan, R.Y., Lui, W.Y., Wu, K., Chan, C.L., Dai, S.H., Sui, Z.Q., Corke, H., 2017. Bioactive compounds and bioactivities of germinated edible seeds and sprouts: an updated review. Trends in Food Science & Technology 59, 1–14.

García, C., Rendueles, M., Díaz, M., 2019. Liquid-phase food fermentations with microbial consortia involving lactic acid bacteria: a review. Food Research International 119, 207–220.

Gawande, V., Babar, K.,P., 2018. Effect of malting & fermentation on the functional & rheological properties of sorghum flour. Asian Food Science Journal 1, 1–4.

Gawlik-Dziki, U., Uwieca, M., Dziki, D., Seczyk, Ł., Złotek, U., Różyło, R., Kinga, K., Ryszawy, D., Czyż, J., 2014. Anticancer and antioxidant activity of bread enriched with broccoli sprouts. BioMed Research International 2014, 1–14.

Giacalone, D., Degn, T.K., Yang, N., Chujiao, L., Fisk, J., Munchow, M., 2019. Common roasting defects in coffee: aroma composition, sensory characterization and consumer perception. Food Quality and Preference 71, 463–474.

Gibson, R.S., Yeudall, F., Drost, N., Mitimuni, B., Cullinan, T., 1998. Dietary interventions to prevent zinc deficiency. American Journal of Clinical Nutrition 68, 484–487.

Gracka, A., Jeleń, H.,H., Majcher, M., Siger, A., Kaczmarek, A., 2016. Flavoromics approach in monitoring changes in volatile compounds of virgin rapeseed oil caused by seed roasting. Journal of Chromatography A 1428, 292–304.

Guerrini, L., Parenti, O., Angeloni, G., Zanoni, B., 2019. The bread making process of ancient wheat: a semi-structured interview to bakers. Journal of Cereal Science 87, 9–17.

Gupta, M., Abu-Ghannam, N., Gallaghar, E., 2010. Barley for Brewing: characteristic changes during malting, brewing and applications of its by-products. Comprehensive Reviews in Food Science and Food Safety 9, 318–328.

Harasym, J., Bogacz-Radomska, L., Oledzki, R., 2016. Postep cywilizacyjny w produkcji żywności – od żywności tradycyjnej do funkcjonalnej. In: Postep Cywilizacyjny – Stan Obecny I Perspektywy, Tygiel, pp. 106–129.

Hingade, S.,T., Chavan, V.,R., Machewad, G.,M., Deshpande, H.,W., 2019. Studies on effect of malting on physiochemical characteristics of wheat malt and barley malt used for preparation of probiotic beverage. Journal of Pharmacognosy and Phytochemistry 8, 1811–1813.

Hinneh, M., Abotsi, E.E., Van de Walle, D., Tzompa-Sosa, D.A., De Winne, A., Simonis, J., Messens, K., Van Durme, J., Afoakwa, E.O., De Cooman, L., Dewettinck, K., 2019. Pod storage with roasting: a tool to diversifying the flavor profiles of dark chocolates produced from 'bulk' cocoa beans? (part I: aroma profiling of chocolates). Food Research International 119, 84–98.

Hochkogler, C.M., Schweiger, K., Rust, P., Pignitter, M., Rathmayr, J., Bayer, S., Chmelirsch, C., Hüller, L., Marko, D., Lang, R., Hofmann, T., Kurz, A.C., Bytof, G., Lantz, I., Schipp, D., Somoza, V., 2019. Daily consumption of a dark-roast coffee for eight weeks improved plasma oxidized LDL and alpha-tocopherol status: a randomized, controlled human intervention study. Journal of Functional Foods 56, 40–48.

Hou, F., Mu, T., Ma, M., Blecker, C., 2018. Optimization of processing technology using response surface methodology and physicochemical properties of roasted sweet potato. Food Chemistry 278, 136–143.

Hurrell, R., Ranum, P., de Pee, S., Biebinger, R., Hulthen, L., John-son, Q., Lynch, S., 2010. Revised recommendations for iron fortification of wheat flour and an evaluation of the expected impact of current national wheat flour fortification programs. Food and Nutrition Bulletin 31 (Suppl. 1), S7.

Jan, K., Ahmad, M., Rehman, S., Gani, A., Khaqan, K., 2019. Effect of roasting on physicochemical and antioxidant properties of kalonji (*Nigella sativa*) seed flour. Food Measure. https://doi.org/10.1007/s11694-019-00052-4.

Jeong, H., Kim, D., Song, H.,J., Lee, S., Kim, K., Kim, K.-J., Han, G.D., 2019. β-Glucan-enriched fermented barley bran (*Sigumjang meju*) extracts attenuates gastric mucosal injury induced by acute alcohol intake in vivo. Food Bioscience 28, 20–26.

Jungbauer, A., Medjakovic, S., Maturitas, 2012. Anti-inflammatory properties of culinary herbs and spices that ameliorate the effects of metabolic syndrome. Maturitas 71, 227–239.

Kazemipoor, M., Cordell, G.,A., 2015. Clinical effects of caraway, a traditional medicine for weight loss. In: Mukherjee, P.K. (Ed.), Evidence-Based Validation of Herbal Medicine. Elsevier, pp. 339–362.

Khazdair, M.R., Anaeigoudari, A., Hashemzehi, M., Mohebbati, R., 2019. Neuroprotective potency of some spice herbs, a literature review. Journal of Traditional and Complementary Medicine 9, 98–105.

Khusro, A., Aarti, C., Barbabosa-Pilego, A., Rojas Hernández, S., 2019. Anti-pathogenic, antibiofilm, and technological properties of fermented food associated *Staphylococcus succinus* strain AAS2. Preparative Biochemistry & Biotechnology 49, 176–183.

Kimura, K., Yokoyama, S., 2019. Trends in the application of Bacillus in fermented foods. Current Opinion in Biotechnology 56, 36–42.

Kondeva-Burdina, M., Georgieva, M., Kasabova-Angelova, A., Tzankova, V., Zlatkov, A., 2019. Preliminary in vitro evaluation of neuroprotective and monoamine oxidase type B inhibitory effects of newly synthesized 8-Aminocaffeines. Neural Regeneration Research 14, 971–972.

Kudełka, W., 2011. Innowacyjny segment żywności wspierającej zdrowie człowieka. Nierówności społeczne a wzrost gospodarczy 18, 290–302.

Kulkarni, K.,D., Kulkarni, D.,N., Ingle, U.,M., 1991. Sorghum malt-based weaning food formulations: preparation, functional properties, and nutritive value. Food and Nutrition Bulletin 13, 322–327.

Lawson, L.,D., Hunsaker, S.,M., 2018. Allicin bioavailability and bioequivalence from garlic supplements and garlic foods. Nutrients 10, 812.

Leatherhead food research report, 2016. Future Directions for the Global Functional Foods Market.

Li, J., Zhou, R.-L., Ren, Z.-Q., Fan, Y.-W., Hu, S.-B., Zhuo, C.-F., Deng, Z.-Y., 2019. Improvement of protein quality and degradation of allergen in soybean meal fermented by Neurospora crassa. LWT – Food Science and Technology 101, 220–228.

Liu, F., He, C., Wang, L., Wang, M., 2018. Effect of milling method on the chemical composition and antioxidant capacity of Tartary buckwheat flour. International Journal of Food Science and Technology 53, 2457–2464.

Lopez, H.,W., Krespine, V., Guy, C., Messager, A., Demigne, C., Remesy, C., 2001. Prolonged fermentation of whole wheat sourdough reduces phytate level and increases soluble magnesium. Journal of Agricultural and Food Chemistry 49, 2657–2662.

Mallikarjuna, S.,E., Dharmesh, S.,M., 2018. Swallow root (*Decalepis hamiltonii*) pectic oligosaccharide (SRO1) induces cancer cell death via modulation of galectin-3 and survivin. Carbohydrate Polymers 15, 402–410.

Masi, C., Dinnella, C., Barnabà, M., Navarini, L., Monteleone, E., 2013. Sensory properties of under-roasted coffee beverages. Journal of Food Science 78, 1290–1300.

Meng, F.-C., Zhou, Y.-Q., Ren, D., Wang, R., Zhang, Q.-W., 2018. Turmeric: a review of its chemical composition, quality control, bioactivity, and pharmaceutical application. In: Grumezescu, A.M., Holban, A.M. (Eds.), Natural and Artificial Flavoring Agents and Food Dyes. Academic Press, pp. 299–350.

Mertz, D.P., 2008. Civilization diseases: a historical review. Medizine Welt 59 (3–4), 113–115.

Miller Jones, J., Adams, J., Harriman, C., Miller, C., Van der Kamp, J.,W., 2015. Nutritional impacts of different whole grain milling techniques: a review of milling practices and existing data. Cereal Foods World 60, 130–139.

Molinari, R., Costantini, L., Timperio, A.,M., Lelli, V., Bonafaccia, F., Bonafaccia, G., Merendino, N., 2018. Tartary buckwheat malt as ingredient of gluten-free cookies. Journal of Cereal Science 80, 37–43.

Montemurro, M., Pontonio, E., Gobbetti, M., Rizzello, C.G., 2019. Investigation of the nutritional, functional and technological effects of the sourdough fermentation of sprouted flours. International Journal of Food Microbiology 302, 47–58.

Morikawa, H., Okuda, K., Kunihira, Y., Inada, A., Miyagi, C., Matsuo, Y., Saito, Y., Tanaka, T., 2019. Oligomerization mechanism of tea catechins during tea roasting. Food Chemistry 285, 252–259.

Nelson, K., Stojanovska, L., Vasiljevic, T., Mathai, M., 2013. Germinated grains: a superior whole-grain functional food? Canadian Journal of Physiology and Pharmacology 91, 429–441.

Nkhata, S.,G., Ayua, E., Kamau, E.,H., Shingiro, J.,B., 2018. Fermentation and germination improve value of cereals and legumes trough activation of endogenous enzymes. Food Sciences and Nutrition 6, 2446–2458.

Özcan, M.,M., Aljuhaimi, F., Uslu, N., 2018. Effect of malt process steps on bioactive properties and fatty acid composition of barley, green malt and malt grains. Journal of Food Science & Technology 55, 226–232.

Peñaloza-Espinosa, J., De La Rosa-Angulo, G., Mora-Escobedo, R., Chanona-Pérez, J., Farrera-Rebollo, R., Calderón-Domínguez, G., 2011. Sourdough and bread properties as affected by soybean protein addition. In: Ng, T.-B. (Ed.), Soybean – Applications and Technology. InTech, pp. 387–402.

Pérez-Burillo, S., Mehta, T., Pastoriza, S., Kramer, D.,L., Paliy, O., Rufián-Henares, J.,A., 2019. Potential probiotic salami with dietary fiber modulates antioxidant capacity, short chain fatty acid production and gut microbiota community structure. LWT –Food Science and Technology 105, 355–362.

Pouliot, Y., Conway, V., Leclerc, P.L., 2014. Separation and concentration technologies in food processing. In: Clark, S., Jung, S., Lamsal, B. (Eds.), Food Processing: Principles and Applications, second ed. John WIley & Sons, pp. 33–60.

Reale, A., Konietzny, U., Coppola, R., Sorrentino, E., Greiner, R., 2007. The importance of lactic acid bacteria for phytate degradation during cereal dough fermentation. Journal of Agricultural and Food Chemistry 55, 2993–2997.

Rezac, S., Kok, C.,R., Heermann, M., Hutkins, R., 2018. Fermented foods as a dietary source of live organisms. Frontiers in Microbiology 9, 1785.

Rizzello, C.,G., Calasso, M., Campanella, D., De Angelis, M., Gobbetti, M., 2014. Use of sourdough fermentation and mixture of wheat chickpea, lentil and bean flours for enhancing the nutritional, texture and sensory characteristics of white bread. International Journal of Food Microbiology 180, 78–87.

Rizzello, C.,G., Coda, R., Gobbetti, M., 2017. Use of sourdough fermentation and nonwheat flours for enhancing nutritional and healthy properties of wheat-based foods. In: Frias, J., Martinez-Villaluenga, C., Peñas, E. (Eds.), Fermented Foods in Health and Disease Prevention. Academic Press, Boston, FL, USA, pp. 433–452.

Rocchetti, G., Miragoli, F., Zacconi, C., Lucini, L., Rebecchi, A., 2019. Impact of cooking and fermentation by lactic acid bacteria on phenolic profile of quinoa and buckwheat seeds. Food Research International 119, 886–894.

Rocha, I.S., de Santana, L.R.R., Soares, S.E., da Silva Bispo, E., 2017. Effect of the roasting temperature and time of cocoa beans on the sensory characteristics and acceptability of chocolate. Food Science and Technology 37. https://doi.org/10.1590/1678-457x.16416.

Rusydi, M.R., Noraliza, C.W., Azrina, A., Zulkhairi, A., 2011. Nutritional changes in germinated legumes and rice varieties. International Food Research Journal 18, 688–696.

Salehi, B., Stojanović-Radić, Z., Matejić, J., Sharifi-Rad, M., Sharifi-Rad, J., 2019. The therapeutic potential of curcumin: a review of clinical trials. European Journal of Medicinal Chemistry 163, 527–545.

Schlemmer, U., Frølich, W., Prieto, R.,M., Grases, F., 2009. Phytate in foods and significance for humans: food sources, intake, processing, bioavailability. Molecular Nutrition & Food Research 53, 330–375.

Schlörmann, W., Birringer, M., Böhm, V., Löber, K., Jahreis, G., Lorkowski, S., Müller, A.,K., Schöne, F., Glei, M., 2015. Influence of roasting conditions on health-related compounds in different nuts. Food Chemistry 180, 77–85.

Seibel, J., Kryshen, K., Pongrácz, J.E., Lehne, r M.D., 2018. In vivo and in vitro investigation of anti-inflammatory and mucus-regulatory activities of a fixed combination of thyme and primula extracts. Pulmonary Pharmacology & Therapeutics 51, 10–17.

Shakeri, A., Cicero, A.F.G., Panahi, Y., Mohajeri, M., Sahebkar, A., 2019. Curcumin: a naturally occurring autophagy modulator. Journal of Cellular Physiology 234, 5643–5654.

Sharma, A., Singh, J., Kumar, S., 2012. Bay leaves. In: Peter, K.,V. (Ed.), Handbook of Herbs and Spices, second ed. Woodhead Publishing, pp. 73–85.

Shi, Q., Hou, Y., Xu, Y., Krogh, K. B.,R.,M., Tenkanen, M., 2019. Enzymatic analysis of levan produced by lactic acid bacteria in fermented doughs. Carbohydrate Polymers 208, 285–293.

Sivamaruthi, B., Kesika, P., Chaiyasut, C., 2018. Impact of fermented foods on human cognitive function—a review of outcome of clinical trials. Scientia Pharmaceutica 86, 22.

Średnicka-Tober, D., Barański, M., Seal, C., Sanderson, R., Benbrook, C., Steinshamn, H., Gromadzka-Ostrowska, J., Rembiałkowska, E., Skwarło-Sońta, K., Eyre, M., Cozzi, G., Krogh Larsen, M., Jordon, T., Niggli, U., Sakowski, T., Calder, P.,C., Burdge, G.,C., Sotiraki, S., Stefanakis, A., Yolcu, H., Stergiadis, S., Chatzidimitriou, E., Butler, G., Stewart, G., Leifert, C., 2016. Composition differences between organic and conventional meat: a systematic literature review and meta-analysis. British Journal of Nutrition 115, 994–1011.

Srinivasan, K., 2017. Ginger rhizomes (*Zingiber officinale*): a spice with multiple health beneficial potentials. PharmaNutrition 5, 18–28.

Suleria, H.A.R., Butt, M.S., Khalid, N., Sultan, S., Abbas, M., 2015. Garlic (*Allium sativum*): diet based therapy of 21st century—a review. Asian Pacific Journal of Tropical Disease 5, 271–278.

Tamang, J.,P., Watanabe, K., Holzapfel, W.,H., 2016. Diversity of microorganisms in global fermented foods and beverages. Frontiers in Microbiology 7, 377.

Van Asselt, E.,D., Capuano, E., Van der Fels-Klerx, H.,J., 2015. Sustainability of milk production in the Netherlands: a comparison between raw organic, pasteurized organic and conventional milk. International Dairy Journal 47, 19–26.

Wichienchot, S., Ishak, W.R., 2017. Sustainability of nutraceuticals and functional foods. In: Bhat, R. (Ed.), Sustainability Challenges in the Agrofood Sector. John Wiley & Sons Ltd., pp. 369–389. https://doi.org/10.1002/9781119072737.ch16

Wongsa, P., Khampa, N., Horadee, S., Chaiwarith, J., Rattanapanone, N., 2019. Quality and bioactive compounds of blends of Arabica and Robusta spray-dried coffee. Food Chemistry 283, 579–587.

Xu, F.-H., Gao, Q.-H., Ma, Y.-J., Guo, H.-D., Wang, M., 2014. Comparison of tartary buckwheat flour and sprouts steamed bread in quality and antioxidant property. Journal of Food Quality 37, 318–328.

Xue, Y.-F., Zhang, W., Liu, D.-Y., Xia, H.-Y., Zou, C.-Q., 2016. Nutritional composition of iron, zinc, calcium, and phosphorus in wheat grain milling fractions as affected by fertilizer nitrogen supply. Cereal Chemistry 93, 543–549.

Yang, N., Liu, C., Liu, X., Degn, T.,K., Munchow, M., Fisk, I., 2016. Determination of volatile marker compounds of common coffee roast defects. Food Chemistry 211, 206–214.

Ye, J.-H., Huang, L.-Y., Terefe, N.S., Augustin, M.A., 2019. Fermentation-based biotransformation of glucosinolates, phenolics and sugars in retorted broccoli puree by lactic acid bacteria. Food Chemistry 286, 616–623.

CHAPTER

Effect of nonthermal technologies on functional food compounds

8

Silvia Tappi[1], Urszula Tylewicz[1], Marco Dalla Rosa, PhD [1,2]

[1]*Department of Agricultural and Food Sciences, Alma Mater Studiorum-Università di Bologna, Cesena, Forlì - Cesena, Italy;* [2]*Professor, Department of Agricultural and Food Sciences, Alma Mater Studiorum-Università di Bologna, Cesena, Forlì - Cesena, Italy*

1. Introduction

In the past decades, many emerging, nonthermal technologies have been proposed as a potential alternative to the conventional thermal treatments to overcome limitations due to the use of heat and to improve quality of the product and hence the competitiveness on the market. Moreover, many of these techniques have shown good potentiality in reducing the environmental impact of food processing and increasing the sustainability of the food chain by improving the recovery of high added value components from food waste and by-products. Among these techniques, pulsed electric fields (PEF), ultrasound (US), high hydrostatic pressure (HHP), and cold plasma (CP) showed to be very promising tools for the sustainable novel food production, since they require lower energy consumption and minimizing resource inputs. Moreover, these technologies could be used for the improvement of food quality and safety.

The effect of these technologies on functional components of foods has been widely investigated. PEF and HHP are considered often alternative to the traditional thermal treatments and generally allowed to better maintain the thermolabile compounds such as vitamins. Moreover, PEF, HHP, and US affecting the integrity of cell membranes allow to increase mass transfer phenomena, thus decreasing processing times during, for example, osmotic dehydration (OD), drying, freezing etc., but also to extract functional components from food by-products, thus contributing to their valorization.

CP is an emerging technology for decontamination of food and packaging surfaces, which allow to operate at room temperature, with reduced energy requirements and short processing times. However. being highly oxidative can be detrimental for various food components, such as vitamins and polyphenols; however, the changes in nutritional properties seemed to be only minor and depended strongly on food matrixes and the parameter used.

This chapter provides a description of the main application of proposed technologies in food industry in relation to the food sustainable production, as well as their

Sustainability of the Food System. https://doi.org/10.1016/B978-0-12-818293-2.00008-2

effect on the bioactive compounds extraction, retention, or degradation. The recovery of high-added value ingredients and developing of new products can significantly improve the sustainability of the food production system.

2. Pulsed electric field

The phenomenon of cell membrane electroporation occurs when biological cells are exposed to an external electric field (E) by application of short electric pulses, with an intensity higher than a threshold value (Weaver and Chizmadzhev, 1996). This phenomenon is associated with the membrane permeability and leads to the formation of transient or permanent pores in the membrane that consequently compromises its semipermeability (Donsì et al., 2010). Transient pores (called also reversible) are created in the membrane when a low E is applied; after removal of the electrical field, the membrane functionality is restored. When high E is used, the irreversible tissue permeabilization occurs (permanent pores creation and membrane damage), which lead to the cell death (Teissie et al., 2005; Tylewicz et al., 2017; Montie et al., 2000).

2.1 Applications of pulsed electric field

PEF technology has been widely studied for both food preservation and processing. Irreversible electroporation is mainly used to inactivate enzymes (Elez-Martínez et al., 2006) and microorganisms (Sánchez-Moreno et al., 2005). PEF processing, especially at low and medium intensities, offers several advantages in processes related to the mass transfer phenomena, i.e., improving extraction yield (Barba et al., 2015a) and reducing time and saving energy in processes such as OD, drying, freeze-drying, and freezing (Donsì et al., 2010; Wiktor et al., 2015; Barba et al., 2015a; Tylewicz et al., 2016, 2017). Therefore, the potentiality of PEF applications in the food industry is huge, and it has been revised by Barba et al. (2015a). Moreover, recently, PEF has received increasing attention, because of the possibility for the manipulation of biological cells and tissues; in fact, reversible electroporation could be used to incorporate different functional substances or drugs into the biological tissue (Gómez Galindo, 2017; Poojary et al., 2017).

Considering literature reports in the past decade, the contribution of PEF to sustainability is related to the improved process efficiency (higher extraction yield, reduced extraction time, temperature, and use of solvents), the improvement of further purification steps, and the reduction of energy requirements, costs, and environmental impact (Barba et al., 2015a).

2.2 Effect on functional components

Several studies have been conducted evaluating the impact of PEF on bioactive compounds retention and/or extraction. When used as alternative to thermal pasteurization, PEF treatment was shown to better preserve bioactive components such as vitamin C in apple juice (Dziadek et al., 2019), in mango nectar (Kumar et al., 2019), and in orange juice (Elez-Martínez et al., 2006). Similar results were observed for total polyphenol content. Some authors observed the preservation of flavanone content in orange juice (Sánchez-Moreno et al., 2005) upon the application of 35 kV/cm (750- and 4-μs bipolar pulses at 800 Hz) and total isoflavone content in fruit juice-soymilk beverage (Morales-De La Peña et al., 2010) treated at 35 kV/cm (800 and 1400 μs), while an increase in the phenolic content, mainly hesperidin, of a PEF-treated (35 kV/cm, 4-μs bipolar pulses at 200 Hz, for 800 or 1400 μs) fruit juice-soymilk beverage was observed by Morales-De La Peña et al. (2011).

The application of PEF was also shown to increase the extraction efficiency of bioactive compounds from different by-products and waste of the food processing line, such as orange peels (Luengo et al., 2013), grape by-products (Corrales et al., 2008; Boussetta et al., 2012), fresh tea leaves (Liu et al., 2019), blackberries (Barba et al., 2015b), and blueberries and their by-products (Pataro et al., 2017). Table 8.1 shows some of the many literature reports related to effect of PEF on extraction of functional components from food by-products.

Pataro et al. (2018) applied PEF at low electric field strength (<1 kV/cm) combined with steam blanching (SB) on tomatoes, for the reduction of the energy required for peeling. The main aim was to increase the efficiency of the steam-peeling process; however, as a side effect, the authors observed an increased extractability of carotenoids, in particular lycopene, from the waste demonstrating that the integration of PEF in the processing line of tomato fruits prior to SB contributes to the valorization of tomato processing by-products. Similar results were observed by Odriozola-Serrano et al. (2007) for lycopene and by Kumar et al. (2019) and Ade-Omowaye et al. (2002) for total carotenoids.

Barbosa-Pereira et al. (2018) used PEF to obtain high added value from coffee silverskin and cocoa bean shells. Results showed that extraction was significantly increased compared with the traditional methods, indicating these by-products as potential new sources for the recovery of compounds with beneficial effects and with several applications in the food (as food ingredients and nutraceuticals), cosmetic, and pharmaceutical industries.

Ferreira et al. (2019) conducted an life cycle assessment (LCA) study on the production of olive oil and wine from PEF-treated products. In both processing lines, a reduction of maceration time was obtained, allowing to reduce energy requirements, and for olives, a 5% increase in the oil yield was also observed, showing how the application of PEF can be implemented to reduce environmental impact of the food production system.

Moreover, PEF has also been investigated for the valorization of by-products of animal origin. Meat and fish processing lead to the production of high amount of waste and by-products, posing big issues of disposal. Many of the by-products contain huge amount of high value bioactive compounds, the recovery of which could highly increase the sustainability of the production process.

Table 8.1 Effect of pulsed electric fields on extraction of functional components from food by-products.

Product	Process parameters	Results	References
Orange peel	5–50 pulses of 3 μs (15–150 μs) E = 1–7 kV/cm Specific energy: 0.06–3.77 kJ/kg Frequency: 1 Hz	Increments in extraction yield and antioxidant activity up to 159% and 192%, respectively	Luengo et al. (2013)
Tomato fruit and peels	E = 0.25–0.75 kV/cm total specific energy 1 kJ/k) Frequency: 10 Hz Pulse width: 20 μs	Improvement of peeling efficiency, increased extraction of carotenoids	Pataro et al. (2018)
Grape by-products	E = 3 kV/cm 30 pulses, specific energy: 10 kJ/kg Frequency: 2 Hz Treatment time: 15 s	Fourfold increase of antioxidant activity of extract compared with traditional extraction Higher recovery of selected polyphenols	Corrales et al. (2008)
Grape seeds	E = 8–208 kV/cm n_{PEF} = 2000 pulses Frequency f: 0.33 Hz Total time: 1–20 ms	Increased polyphenol extraction	Boussetta et al. (2012)
Cocoa and coffee bean products	E = 1.5–4.4 kV/cm Pulses n = 500–1000 Treatment time: 5–20 μs	Extractability of bioactive compounds from cocoa and coffee by-products improved by 20% and 21.3%, respectively, compared with untreated sample	Barbosa-Pereira et al. (2018)
Olives for oil production	E = 2.5 kV/cm, specific energy 3.4 kJ/kg Pulse width: 20 μs	Decrease of maceration time: reduction of environmental impact between 3% and 14% 5% increase of oil extraction reduced the environmental impact indicators approximately 5%	Ferreira et al. (2019)
Grape for wine making	E = 6 kV/cm, energy input: 8.3 kJ/kg Pulse width: 100 μs	Decrease of maceration time: reduction of environmental impact between 3% and 14%	Ferreira et al. (2019)
Waste chicken breast muscle	Three protocols combining E = 80–400 V/cm Pulse	Extraction of functional molecules from meat waste without use of chemicals	Gosh et al. (2019)

Table 8.1 Effect of pulsed electric fields on extraction of functional components from food by-products.—*cont'd*

Product	Process parameters	Results	References
	width = 50 μs –5 ms Pulse number = 50 –100 Energy value of 8.4 ± 1.2 J/g		
Fishbone	E = 5, 10, 15, 20, and 25 kV/cm Pulse number = 2, 4, 6, 8, 10, and 12	Accelerated extraction and higher yields for chondroitin sulfate, optimized conditions E = 16.88, and pulse n§ = 9	He et al. (2014)

Ghosh et al. (2019) investigated the use of PEF for recovering functional proteins from chicken meat waste and obtained a fraction enriched in proteins with possible antioxidant properties applying a total energy of 38.4 kJ/kg. Similarly, He et al. (2014) extracted chondroitin sulfate from fishbone by high-intensity PEF, highlighting the benefits of this technique including reduced process time, enhanced efficient, and eco-friendly aspects.

3. Ultrasound technology

US is a mechanical wave at a frequency above human hearing, in the range from 16 kHz to 100 MHz. In particular, the high-intensity (power) US operates at the frequency range of kHz (McClements, 1995; Nowacka et al., 2014; Zheng and Sun, 2006). The principle of US application is based on fast compression and expansion of the tissue (sponge effect) and cavitation phenomena. The cavitation occurs due to the generation, growth, and collapse of large bubbles, which causes the liberation of higher energies (McClements, 1995). Both cavitation and compression/expansion of the tissue can cause disruption of cell walls and membranes, which results in the formation of microscopic channels in the biological material, causing the alteration of physical and chemical properties of food products (Nowacka et al., 2017, 2019; Wiktor et al., 2016).

3.1 Applications of power ultrasound

Power US finds many applications in food technology. It can be used to enhance different processes based on heat and mass transfer (e.g., OD, drying, freezing and thawing, extraction, filtration, crystallization, or emulsification). Such a wide range of application of US is associated with the modifications that occur during sonication in both the solid and liquid phase of the food material (Nowacka et al., 2014, 2017; Witrowa-Rajchert et al., 2014). US-assisted extraction is commonly used nowadays in food technology as a substitute for the conventional extraction technique, to improve the recovery of bioactive compounds from plant materials. Moreover, high-power US has been successfully used in food enzymes and microbial inactivation (Inguglia et al., 2018). The application of US in food science and technology has been widely reviewed by Gallo et al. (2018).

3.2 Effect on functional components

US-assisted extraction (UAE) can be used as a pretreatment or during the actual solid-liquid extraction step. UAE is considered novel clean, green extraction technology for various molecules and biomaterials, including polysaccharides, essential oils, proteins, peptides, fine chemicals (dyes and pigments), and bioactive molecules of commercial importance (Tiwari, 2015). The lower requirements for solvent and the reduction of extraction times make UAE environmentally friendly. Moreover, US is relatively easy to use, versatile, and flexible and requires low investment compared with other novel extraction techniques (e.g., supercritical fluid extraction, pressurized solvent extraction). Table 8.2 reports some results obtained by applying UAE for the recovery of functional components from fruit and vegetable by-products.

Virot et al. (2010) investigated the use of UAE for polyphenol extraction from apple pomace, in comparison with conventional maceration procedure. Results showed that the use of US allowed to increase the yield by 20% with a reduced time. Similarly, Pingret et al. (2012) found an increase of 30%–50% in the extraction of polyphenols from the same matrix with 40 min at 0.764 W/cm^2.

For the extraction of polyphenols from apple pomace, Corrales et al. (2008) combined US with heat at 70°C (thermosonication) and observed a 50% increase of the extraction yields compared with conventional extraction. Minjares-Fuentes et al. (2014) optimized the extraction of pectin from grape pomace. They found that by applying 37 kHz and 140 W, a 20% higher pectin yield was obtained compared with traditional extraction. Moreover, it was possible to obtain pectin with higher average molecular weight.

Various authors found that UAE showed a significantly higher efficiency compared with conventional extraction techniques in terms of reduced extraction time, increased yield, and lower temperature (Virot et al., 2010). Papoutsis et al. (2018) used different models to optimize the extraction of polyphenols by US from lemon by-products. Moreover, Roseiro et al. (2013) observed that, by combined UAE with supercritical fluid extraction, the valorization of carob gum by-products was possible to obtain higher yields, reducing the operating temperature and thus reducing operating costs.

The increased sustainability of the extraction process can be obtained also by reducing the use of solvents. Adam et al. (2012) optimized the extraction of oil from microalgae cells using an innovative US technology. With a total solvent-

Table 8.2 Effect of ultrasound on the extraction of functional components from fruit and vegetable by-products.

Product	Process parameters	Results	References
Lemon by-products	150–250 W, 35–45 min, 45–55°C	Maximum total polyphenol content extraction: 18.10 ± 0.24 mg GAE/g_{dw} at US power of 250 W, 45 min, 50°C Maximum rutin extraction: 3.20 ± 0.12 mg/g_{dw} at US power of 150 W, 35 min, 48°C	Papoutsis et al. (2018)
Apple pomace	25 kHz, 150 W, 5–55 min	20% increased catechin extraction compared with traditional method, reduced extraction times	Virot et al. (2010)
Pomegranate seeds	140–180 W, 35–45°C, 20–40 min	Higher yield in oil compared with other technologies	Tian et al. (2013)
Grape pomace	37 kHz, 140 W	20% higher pectin yield compared with traditional extraction. Pectin with higher average molecular weight	Minjares-Fuentesa (2014)
Carob biomass	30°C for 30 min, using a Transsonic T700 sonifier (320 W, 35 kHz	Higher yields reducing the operating temperature and thus reducing operating costs	Roseiro et al. (2013)
Apple pomace	40°C, 40 min, and 0.764 W/cm^2	30% increase of polyphenols and antioxidant activity compared with traditional extraction	Pingret et al. (2012)
Microalgae	1000W, 30 min	Solvent-free extraction of lipids	Adam et al. (2012)

free extraction, it was possible to better preserve the bioactive compounds compared with traditional methods in an industrial scale process.

The use of US has also been applied as pretreatment to different technologies involving mass transfer with the aim of reducing processing times and thus increasing process sustainability. At the same time, the use of US allows a good preservation of functional components compared with the traditional process. Frias et al. (2010) observed a higher vitamin C (82%–92%) retention in US-assisted drying (20°C for 120 min) in carrots compared with the traditional convective air drying.

Moreover, the use of US seems to increase the availability of different functional components and increase the availability of vitamins such as B1, B2, B3, B6, and B5, during ultrasonic-assisted air drying of cherry tomatoes (Fernandes et al.,

2016), or vitamin C in cashew apple bagasse (Fonteles et al., 2017) because of the release of the components from the membrane, protein, or apoenzyme. Hence, beside the reduction of processing times, a higher nutritional quality of products can be obtained by applying US.

4. High hydrostatic pressure

High-pressure processing (HPP), or HHP processing, involves the application of pressures in the range of 100−1000 MPa typically applied and maintained for a short time (1−15 min) and transmitted through water to a food product.

During the HPP process, high pressure is uniformly and instantaneously applied through a food material, independent of its mass, shape, and composition (isostatic principle), causing reactions responsible for the reduction in volume of biomolecules (Le Chatelier principle) (Heremans, 2002; Valdez-Fragoso et al., 2011).

Industrial HHP installations typically operate discontinuously and can attain pressures of up to 800 MPa, although pressures exceeding 400 MPa are not normally used for solid foods because they can bring about a reversible or irreversible disruption of inter- and intramolecular bonds (Knorr et al., 2006; Heinz and Buckow, 2010).

4.1 Application of high hydrostatic pressure

The main application of high pressure is related to the inhibition of microorganisms, although also enzymes can be inactivated (USFDA, 2000; Saucedo-Reyes et al., 2009). The combination of HHP with mild heat treatment (10−40°C) allows to pasteurize products that will require a further refrigerated storage, while, to reach the sterilization effect, however, HHP needs to be combined with elevated temperature (60−90°C) (Mújica-Paz et al., 2011).

HHP treatment has been investigated for the preserving of a wide range of foods, including juices and beverages, fruits, and vegetables (Heinz and Buckow, 2010; Pereira and Vicente, 2010; Mújica-Paz et al., 2011) but also meat, fish, and seafood products (Campus, 2010). Toepfl et al. (2006) attempted to evaluate the energy efficiency of HHP compared with thermal treatments. While a 20% reduction in energy requirement was calculated for sterilization, authors underlined the high investment costs and cost-intensive maintenance and service that limit the industrial application.

Moreover, because of the effect on cell membranes, HHP allows to accelerate mass transfer phenomena and hence shows good potentiality as a pretreatment for processes of drying or dehydration (Luo et al., 2019) with the aim of reducing treatment time. The enhanced cell disruption promoted by HHP can be exploited also for the valorization of by-products through the extraction of bioactive compounds (Corrales et al., 2008).

4.2 Effect on functional compounds

The effect on HHP on nutritional aspects has been studied on a variety of matrices and in relation to different functional components. However, the most investigated component is certainly ascorbic acid or vitamin C. Being highly thermal sensitive, vitamin C is rapidly degraded during thermal treatments and can hence represent an index of product nutritional quality. Table 8.3 reports some of the many reports of the effect of HHP on vitamin C content in different food matrixes compared with the traditional thermal treatment.

Although a decrease of vitamin C could be observed in various food matrixes, when compared with a corresponding thermal treatment, HHP seems to reduce vitamin loss and to improve its maintenance during storage. Patras et al. (2009) observed only 7% loss after 600 MPa treatment for 15 min of tomato purees, compared with the 54% loss following thermal treatment. In orange juice subjected to 400 MPa at 40°C for 1 min, the decrease was 8% lower compared with the thermal treated sample (Plaza et al., 2006). In the same research, the author observed also that HHP allowed to better preserve vitamin C and antioxidant activity during refrigerated storage. Similarly, Jayachandran et al. (2015) obtained a reduction of 3%–27% by applying HHP in the range of 200–600 MPa/30–70°C to a litchi-

Table 8.3 Examples of the effect of high hydrostatic pressure (HHP) on ascorbic acid and vitamin C content of different food products compared with the traditional thermal treatment.

Product	Process parameters	Results	References
Tomato purées	400–600 MPa/20°C/ 15 min	Lower decrease in vitamin C (–7%) after HHP treatments compared with thermal processing (–54%).	Patras et al. (2009)
Tomato purée	400 MPa/25°C/15 min	Vitamin C losses about 29%, similar to thermal treatment	Sánchez-Moreno et al. (2006)
Litchi-based mixed fruit beverage	200–600 MPa/30–70°C/ 0–20 min	Loss of 3%–27% against 42% of pasteurized product	Jayachandran et al. (2015)
Orange juice	400 MPa/40°C/1 min	Loss of vitamin C 8% lower compared with thermal-treated sample. Better maintenance during storage	Plaza et al. (2006)
Green plum	50 MPa/1 min before osmotic dehydration (OD) in 40% sucrose solution for 5 days	No changes after HHP, while heating promoted 17% loss. Better retention during OD process	Luo et al. (2019)

based beverage, whereas 42% was observed after traditional pasteurization. On the contrary, Sánchez-Moreno et al. (2006) did not notice any difference between thermal-treated and HHP-treated tomato purée. The stability of vitamin C during high-pressure treatments depends mainly on the oxygen content due to aerobic degradation pathway (Oey et al., 2008). Generally, losses of vitamin C vary according to the matrix, the presence of endogenous prooxidants as enzymes or metal ions, and the level of applied pressure.

Luo et al. (2019) observed that HHP at 50 MPa for 1 min allowed to maintain ascorbic acid content in green plum while heating reduced it by 17%. Moreover, a higher retention was also observed during 5 days of OD.

With regard to other vitamins, various studies have found that different components belonging to the B group are resistant to HHP in both model systems and food matrixes (Butz et al., 2007; Gabrovska et al., 2005). Folic acid (B9), however, showed high percentages of degradation, depending on the specific folate derivate considered (Oey et al., 2008).

Studies on stability of fat-soluble vitamins upon HPP treatments are less abundant in the literature. However, various reports have observed that high pressure increases the plant cell permeability, facilitating extraction of carotenoids and hence their bioaccessibility (Nagarajan et al., 2016). This increase corresponded, for example, to 25% in orange, papaya, melon, and milk smoothies after 450 and 600 MPa treatments (Andrés et al., 2016) and to 54% in orange juice after 400 MPa treatment at 40°C for 1 min (Sánchez-Moreno et al., 2005).

Increased extraction was also observed for vitamin A in orange juice of 38.74% after 400 MPa at 40°C for 1 min (Sanchez-Moreno et al., 2005) or up to 45% after 350 MPa combined with 30–60°C (De Ancos et al., 2002).

The increased extraction of components was also exploited for the valorization of some food by-products. Naghshineh et al. (2013) observed a higher pectin yield from lime peel treated at 100–200 MPa with addition of pectolytic enzymes compared with the traditional extraction method. Corrales et al. (2008) obtained an increased polyphenols extraction and antioxidant activity from grape by-products treated at 600 MPa at 70°C.

5. Cold plasma

Cold atmospheric plasma technology is an emerging technology that could be applied for decontamination and stabilization in the food industry both on packaging material and on the surface of perishable foods. It is based on the ionization of a gas mixture that leads to the formation of excited molecules, ions, electrons, and radical species that coexist with electromagnetic radiations (UV and visible light) (Misra et al., 2016).

There are different technologies used for the generation of CP described in detail by Montie et al., (2000). Moreover, it is possible to obtain plasma using different gas mixtures and gas flow rates. All these parameters influence the exact composition of

CP in terms of temperature, degree of ionization, electron density, electron temperature, type and concentration of reactive species, and occurring UV radiation (Surowsky et al., 2014).

5.1 Application of cold plasma

The high oxidative power of CP has been demonstrated on various microorganisms and food matrixes (Surowsky et al., 2014). The exposure to plasma has also been proven to change some physicochemical properties of food components such as starches and proteins; however, all the different types of plasma generators and the different operating parameters make the comparison of results difficult.

CP is a versatile technology that has been investigated for many applications at both primary and secondary stages of the food production process, such as the treatment of raw materials, intermediate or finished products, and also of packaging materials, processing equipment, and environment. The contribution to the sustainability of the production relies on the possibility of operating at low temperatures, reduces processing times and energy requirement, and antimicrobial efficacy (Bourke et al., 2018).

5.2 Effect on functional compounds

Recently, attention has been drawn toward the effect of CP on the bioactive content of food products. Indeed, the highly oxidative power of plasma reactive species can represent a detrimental factor for various food components. Table 8.4 reports some of the main findings regarding the effect of CP on ascorbic acid and polyphenol content of a range of food products.

Grzegorzewski et al. (2010, 2011) observed a detrimental effect on the tissue photosynthetic efficiency and a time and structure-dependent degradation of phenolic compounds after CP exposure of green leafy vegetables. However, the oxidation of the bioactive components was mitigated by the matrix effect. Almeida et al. (2015) found no variation in total phenolic content of prebiotic orange juice, whereas Pankaj et al. (2017) observed a decrease in total phenolics and flavonoids of white grape juice, comparable with the one obtained by thermal treatment. Anthocyanins in blueberries were observed to decrease proportionally to treatment times by Lacombe et al. (2015) and Sarangapani et al. (2017). Only a slight decrease in the total phenolic content and antioxidant activity of apple slices was observed. However, treated polyphenol extracts did not reduce cell viability and did not suppress the physiological response to oxidative stress in human-cultured colonocytes (Ramazzina et al., 2016).

On the other side, some authors observed that a short exposition to CP promoted an increase in the content of some specific phenolic compounds in apples (Tappi et al., 2018) and blueberries (Sarangapani et al., 2017). This increase was attributed to a physiological response of the tissue to the stress promoted by the reactive

Table 8.4 Effect of cold atmospheric plasma on bioactive compounds content of different food products.

Product	Process parameters	Results	References
Vitamin C			
Kiwifruit	Dielectric barrier discharge (DBD) plasma generator, potential difference 15 kV, input voltage 19 V Gas: air	No effect on vitamin C content after treatment and after 4 d storage	Ramazzina et al. (2015)
Radish sprouts	Microwave generator (2.45 GHz) Gas: N_2 flow: 1 L/min generation power: 900 W Pressure: 667 Pa 0, 2, 5, 10, and 20 min	Vitamin C content not affected	Oh et al. (2017)
Orange juice	High-voltage cold plasma 90 kV for 30, 60, and 120 s, direct and indirect treatment	Decrease proportional to treatment time	Xu et al. (2017)
Polyphenols			
Apple slices	Dielectric barrier discharge (DBD) plasma generator, potential difference 15 kV, input voltage 19 V Gas: air Treatment time: up to 30 min, 22°C 60% relative humidity (RH)	Only slight decrease of polyphenol content and in vitro antioxidant activity	Ramazzina et al. (2016)
Apple slices	Dielectric barrier discharge (DBD) plasma generator, potential difference 15 kV, input voltage 19 V Gas: air Treatment time: up to 120 min	Increase of phenolic content at the shortest treatment time and then decrease by increasing exposure	Tappi et al. (2018)
Kiwifruit	Dielectric barrier discharge (DBD) plasma generator, potential difference 15 kV, input voltage 19 V Gas: air Treatment time: up to 40 min	No effect after treatment and after 4 d storage	Ramazzina et al. (2015)

Table 8.4 Effect of cold atmospheric plasma on bioactive compounds content of different food products.—*cont'd*

Product	Process parameters	Results	References
Prebiotic orange juice	DBD-ACP, model: 6CP120/60–7.5-Phenix Technologies 70 kV, treatment times: up to 60 s, direct and indirect exposure	No effect on polyphenol content and antioxidant activity	Almeida et al. (2015)
White grape juice	In packaged treatment 80 kV (peak-to-peak) Treatment time: up to 4 min 46% relative humidity, 24°C	Decrease in total phenolics, total flavonoids and antioxidant capacity, but comparable with those resulting from thermal pasteurization	Pankaj et al. (2017)
Blueberries	In packaged treatment 60 and 80 kV Treatment time: up to 5 min	Increase of total phenols and total flavonoid after 1 min, then decrease with increasing treatment time. Decrease of anthocyanins	Sarangapani et al. (2017)
Blueberries	Power: 549 W at frequency of 47 kHz Treatment time: up to 120 s Gas: air	Decrease in anthocyanins proportional to treatment time	Lacombe et al. (2015)

species. However, the increase of exposure time promoted oxidation reaction and the progressive decrease in the phenolic content and antioxidant activity.

Among vitamins, effect of CP has been studied on vitamin C, which was found only slightly affected upon plasma exposure in kiwifruit (Ramazzina et al., 2015) and radish sprouts (Oh et al., 2017). However, a reduction was observed in orange juice (Xu et al., 2017).

6. Conclusions

In conclusion, the described emerging nonthermal technologies contribute to the sustainability of food production in different stages and in different aspects. They all showed higher efficiency compared with the traditional processing methods, due to lower energy requirements and/or reduced processing times. Moreover, US, PEF, and HHP have shown high potentiality in the valorization of by-products through the extraction of bioactive compounds.

A positive contribution to the sustainability of the food process is related to the high quality of the final products, characterized by increased shelf life and high nutritional values. Indeed, the considered emerging technologies allowed to improve the nutritional characteristics of the food products if compared with the correspondent traditional thermal treatment. It appears that the degradation of bioactive components is strictly dependent on the different sensitivities of the singular components and on the matrix, showing the importance of selecting specific treatment times and parameters on the basis of the raw material characteristics and metabolic response.

Hitherto, information on their effect on some components (e.g., vitamin B complex, liposoluble vitamin) is still scarce; further researches are necessary to obtain a more complete knowledge, in particular for the most novel technologies such as CP.

References

Adam, F., Abert-Vian, M., Peltier, G., Chemat, F., 2012. "Solvent-free" ultrasound-assisted extraction of lipids from fresh microalgae cells: a green, clean and scalable process. Bioresource Technology 114, 457–465.

Ade-Omowaye, B.I.O., Rastogi, N.K., Angersbach, A., Knorr, D., 2002. Osmotic dehydration of bell peppers: influence of high intensity electric field pulses and elevated temperature treatment. Journal of Food Engineering 54 (1), 35–43.

Almeida, F.D.L., Cavalcante, R.S., Cullen, P.J., Frias, J.M., Bourke, P., Fernandes, F.A., Rodrigues, S., 2015. Effects of atmospheric cold plasma and ozone on prebiotic orange juice. Innovative Food Science & Emerging Technologies 32, 127–135.

Andrés, V., Villanueva, M.J., Tenorio, M.D., 2016. The effect of high-pressure processing on colour, bioactive compounds, and antioxidant activity in smoothies during refrigerated storage. Food Chemistry 192, 328–335.

Barba, F.J., Parniakov, O., Pereira, S.A., Wiktor, A., Grimi, N., Boussetta, N., Vorobiev, E., 2015a. Current applications and new opportunities for the use of pulsed electric fields in food science and industry. Food Research International 77 (4), 773–798.

Barba, F.J., Galanakis, M.G., Esteve, M.J., Frigola, A., Vorobiev, E., 2015b. Potential use of pulsed electric technologies and ultrasounds to improve the recovery of high-added value compounds from blackberries. Journal of Food Engineering 167 (Part A), 38–44.

Barbosa-Pereira, L., Guglielmetti, A., Zeppa, G., 2018. Pulsed electric field assisted extraction of bioactive compounds from cocoa bean shell and coffee silverskin. Food and Bioprocess Technology 11 (4), 818–835.

Bourke, P., Ziuzina, D., Boehm, D., Cullen, P.J., Keener, K., 2018. The potential of cold plasma for safe and sustainable food production. Trends in Biotechnology 36 (6), 615–626.

Boussetta, N., Vorobiev, E., Le, L.H., Cordin-Falcimaigne, A., Lanoisellé, J.L., 2012. Application of electrical treatments in alcoholic solvent for polyphenols extraction from grape seeds. LWT-Food Science and Technology 46 (1), 127–134.

Butz, P., Bognar, A., Dieterich, S., Tauscher, B., 2007. Effect of high-pressure processing at elevated temperatures on thiamin and riboflavin in pork and model systems. Journal of Agriculture and Food Chemistry 55 (4), 1289–1294.

Campus, M., 2010. High pressure processing of meat, meat products and seafood. Food Engineering Reviews 2 (4), 256–273.

Corrales, M., Toepfl, S., Butz, P., Knorr, D., Tauscher, B., 2008. Extraction of anthocyanins from grape by-products assisted by ultrasonics, high hydrostatic pressure or pulsed electric fields: a comparison. Innovative Food Science & Emerging Technologies 9 (1), 85–91.

De Ancos, B., Sgroppo, S., Plaza, L., Cano, M.P., 2002. Possible nutritional and health-related value promotion in orange juice preserved by high-pressure treatment. Journal of the Science of Food and Agriculture 82 (8), 790–796.

Donsì, F., Ferrari, G., Pataro, G., 2010. Applications of pulsed electric field treatments for the enhancement of mass transfer from vegetable tissue. Food Engineering Reviews 2 (2), 109–130.

Dziadek, K., Kopeć, A., Drózd, T., Kiełbasa, P., Ostafin, M., Bulski, K., Oziembłowski, M., 2019. Effect of pulsed electric field treatment on shelf life and nutritional value of apple juice. Journal of Food Science & Technology 56 (3), 1184–1191.

Elez-Martínez, P., Soliva-Fortuny, R.C., Martín-Belloso, O., 2006. Comparative study on shelf life of orange juice processed by high intensity pulsed electric fields or heat treatment. European Food Research and Technology 222 (3–4), 321–329.

Fernandes, F.A.N., Rodrigues, S., García-Pérez, J.V., Cárcel, J.A., 2016. Effects of ultrasound-assisted air-drying on vitamins and carotenoids of cherry tomatoes. Drying Technology 34 (8), 986–996.

Ferreira, V.J., Arnal, Á.J., Royo, P., García-Armingol, T., López-Sabirón, A.M., Ferreira, G., 2019. Energy and resource efficiency of electroporation-assisted extraction as an emerging technology towards a sustainable bio-economy in the agri-food sector. Journal of Cleaner Production.

Fonteles, T.V., Leite, A.K.F., da Silva, A.R.A., Fernandes, F.A.N., Rodrigues, S., 2017. Sonication effect on bioactive compounds of cashew apple bagasse. Food and Bioprocess Technology 10 (10), 1854–1864.

Frias, J., Peñas, E., Ullate, M., Vidal-Valverde, C., 2010. Influence of drying by convective air dryer or power ultrasound on the vitamin C and β-Carotene content of carrots. Journal of Agricultural and Food Chemistry 58 (19), 10539–10544.

Gabrovska, D., Paulickova, I., Maskova, E., Fiedlerova, V., Kocurova, K., Pruchova, J., et al., 2005. Changes in selected vitamins, microorganism counts and sensory quality during storage of pressurized sprouted seed of alfalfa (*Medicago sativa* L.). Czech Journal of Food Sciences 23 (6), 246e250.

Gallo, M., Ferrara, L., Naviglio, D., 2018. Application of ultrasound in food science and technology: a perspective. Foods 7 (10), 164.

Ghosh, S., Gillis, A., Sheviryov, J., Levkov, K., Golberg, A., 2019. Towards waste meat biorefinery: extraction of proteins from waste chicken meat with non-thermal pulsed electric fields and mechanical pressing. Journal of Cleaner Production 208, 220–231.

Gómez Galindo, F., 2017. Responses of plant cells and tissues to pulsed electric field treatment. Handbook of Electroporation 4, 2621–2635.

Grzegorzewski, F., Rohn, S., Kroh, L.W., Geyer, M., Schlüter, O., 2010. Surface morphology and chemical composition of lamb's lettuce (*Valerianella locusta*) after exposure to a low-pressure oxygen plasma. Food Chemistry 122 (4), 1145–1152.

Grzegorzewski, F., Ehlbeck, J., Schlüter, O., Kroh, L.W., Rohn, S., 2011. Treating lamb's lettuce with a cold plasma–influence of atmospheric pressure Ar plasma immanent species on the phenolic profile of *Valerianella locusta*. LWT-Food Science & Technology 44 (10), 2285–2289.

He, G., Yin, Y., Yan, X., Yu, Q., 2014. Optimisation extraction of chondroitin sulfate from fish bone by high intensity pulsed electric fields. Food Chemistry 164, 205–210.

Heinz, V., Buckow, R., 2010. Food preservation by high pressure. Journal für Verbraucherschutz und Lebensmittelsicherheit 5 (1), 73–81.

Heremans, K., 2002. High pressure bioscience and biotechnology: a century and a decade perspective. In: Hayashi, R. (Ed.), Trends in High Pressure Bioscience and Biotechnology. Elsevier Science, Amsterdam, pp. 1–6.

Inguglia, E.S., Tiwari, B.K., Kerry, J.P., Burgess, C.M., 2018. Effects of high intensity ultrasound on the inactivation profiles of *Escherichia coli* K12 and *Listeria innocua* with salt and salt replacers. Ultrasonics Sonochemistry 48, 492–498.

Jayachandran, L.E., Chakraborty, S., Rao, P.S., 2015. Effect of high pressure processing on physicochemical properties and bioactive compounds in litchi based mixed fruit beverage. Innovative Food Science & Emerging Technologies 28, 1–9.

Knorr, D., Heinz, V., Buckow, R., 2006. High pressure application for food biopolymers. Biochimica et Biophysica Acta (BBA) – Proteins & Proteomics 1764 (3), 619–631.

Kumar, R., Vijayalakshmi, S., Rajeshwara, R., Sunny, K., Nadanasabapathi, S., 2019. Effect of storage on thermal, pulsed electric field and combination processed mango nectar. Journal of Food Measurement and Characterization 13 (1), 131–143.

Lacombe, A., Niemira, B.A., Gurtler, J.B., Fan, X., Sites, J., Boyd, G., Chen, H., 2015. Atmospheric cold plasma inactivation of aerobic microorganisms on blueberries and effects on quality attributes. Food Microbiology 46, 479–484.

Liu, Z., Esveld, E., Vincken, J.P., Bruins, M.E., 2019. Pulsed electric field as an alternative pre-treatment for drying to enhance polyphenol extraction from fresh tea leaves. Food and Bioprocess Technology 12 (1), 183–192.

Luengo, E., Álvarez, I., Raso, J., 2013. Improving the processing extraction of polyphenols of orange peel by pulsed electric fields. Innovative Food Science & Emerging Technologies 17, 79–84.

Luo, W., Tappi, S., Wang, C., Yu, Y., Zhu, S., Dalla Rosa, M., Rocculi, P., 2019. Effect of high hydrostatic pressure (HHP) on the antioxidant and volatile properties of candied *Wumei* fruit (*Prunus mume*) during osmotic dehydration. Food and Bioprocess Technology 12 (1), 98–109.

McClements, D.J., 1995. Advances in the application of ultrasound. Trends in Food Science & Technology 6, 293–299.

Minjares-Fuentes, R., Femenia, A., Garau, M.C., Meza-Velázquez, J.A., Simal, S., Rosselló, C., 2014. Ultrasound-assisted extraction of pectins from grape pomace using citric acid: a response surface methodology approach. Carbohydrate Polymers 106, 179–189.

Misra, N.N., Schlüter, O., Cullen, P.J. (Eds.), 2016. Cold Plasma in Food and Agriculture: Fundamentals and Applications. Academic Press.

Montie, T.C., Kelly-Wintenberg, K., Roth, J.R., 2000. An overview of research using the one atmosphere uniform glow discharge plasma (OAUGDP) for sterilization of surfaces and materials. IEEE Transactions on Plasma Science 28 (1), 41–50.

Morales-De La Peña, M., Salvia-Trujillo, L., Rojas-Graü, M., Martín-Belloso, O., 2010. Isoflavone profile of a high intensity pulsed electric field or thermally treated fruit juice–soymilk beverage stored under refrigeration. Innovative Food Science & Emerging Technologies 11 (4), 604–610.

Morales-De La Peña, M., Salvia-Trujillo, L., Rojas-Graü, M., Martín-Belloso, O., 2011. Changes on phenolic and carotenoid composition of high intensity pulsed electric field and thermally treated fruit juice—soymilk beverages during refrigerated storage. Food Chemistry 129 (3), 982—990.

Mújica-Paz, H., Valdez-Fragoso, A., Tonello Samson, C., Welti-Chanes, J., Torres, J.A., 2011. High-pressure processing technologies for the pasteurization and sterilization of foods. Food and Bioprocess Technology 4, 969—985.

Nagarajan, J., Nagasundara Ramanan, R., Eshwaraiah Raghunandan, M., Galanakis, C.M., Prasad Krishnamurthy, N., 2016. Carotenoids. In: Galanakis, C.M. (Ed.), Nutraceutical and Functional Food Components: Effects of Innovative Processing Techniques. Academic Press.

Naghshineh, M., Olsen, K., Georgiou, C.A., 2013. Sustainable production of pectin from lime peel by high hydrostatic pressure treatment. Food Chemistry 136 (2), 472—478.

Nowacka, M., Tylewicz, U., Laghi, L., Dalla Rosa, M., Witrowa-Rajchert, D., 2014. Effect of ultrasound treatment on the water state in kiwifruit during osmotic dehydration. Food Chemistry 144, 18—25.

Nowacka, M., Tylewicz, U., Romani, S., Dalla Rosa, M., Witrowa-Rajchert, D., 2017. Influence of ultrasound-assisted osmotic dehydration on the main quality parameters of kiwifruit. Innovative Food Science & Emerging Technologies 41, 71—78.

Nowacka, M., Wiktor, A., Anuszewska, A., Dadan, M., Rybak, K., Witrowa-Rajchert, D., 2019. The application of unconventional technologies as pulsed electric field, ultrasound and microwave-vacuum drying in the production of dried cranberry snacks. Ultrasonics Sonochemistry 56, 1—13.

Odriozola-Serrano, I., Aguiló-Aguayo, I., Soliva-Fortuny, R., Gimeno-Añó, V., Martín-Belloso, O., 2007. Lycopene, vitamin C, and antioxidant capacity of tomato juice as affected by high-intensity pulsed electric fields critical parameters. Journal of Agricultural and Food Chemistry 55, 9036—9042.

Oey, I., Van der Plancken, I., Van Loey, A., Hendrickx, M., 2008. Does high pressure processing influence nutritional aspects of plant based food systems? Trends in Food Science & Technology 19 (6), 300—308.

Oh, Y.J., Song, A.Y., Min, S.C., 2017. Inhibition of *Salmonella typhimurium* on radish sprouts using nitrogen-cold plasma. International Journal of Food Microbiology 249, 66—71, 2017.

Pankaj, S.K., Wan, Z., Colonna, W., Keener, K.M., 2017. Effect of high voltage atmospheric cold plasma on white grape juice quality. Journal of the Science of Food and Agriculture 97 (12), 4016—4021.

Papoutsis, K., Pristijono, P., Golding, J.B., Stathopoulos, C.E., Bowyer, M.C., Scarlett, C.J., Vuong, Q.V., 2018. Optimizing a sustainable ultrasound-assisted extraction method for the recovery of polyphenols from lemon by-products: comparison with hot water and organic solvent extractions. European Food Research and Technology 244, 1353—1365.

Pataro, G., Bobinaitė, R., Bobinas, Č., Šatkauskas, S., Raudonis, R., Visockis, M., Ferrari, G., Viškelis, P., 2017. Improving the extraction of juice and anthocyanins from blueberry fruits and their by-products by application of pulsed electric fields. Food and Bioprocess Technology 10 (9), 1595—1605.

Pataro, G., Carullo, D., Bakar Siddique, MdA., Falcone, M., Donsì, F., Ferrari, G., 2018. Improved extractability of carotenoids from tomato peels as side benefits of PEF treatment of tomato fruit for more energy-efficient steam-assisted peeling. Journal of Food Engineering 233, 65—73.

Patras, A., Brunton, N., Da Pieve, S., Butler, F., Downey, G., 2009. Effect of thermal and high pressure processing on antioxidant activity and instrumental colour of tomato and carrot purées. Innovative Food Science & Emerging Technologies 10, 16—22.

Pereira, R.N., Vicente, A.A., 2010. Environmental impact of novel thermal and non-thermal technologies in food processing. Food Research International 43 (7), 1936–1943.

Pingret, D., Fabiano-Tixier, A.S., Le Bourvellec, C., Renard, C.M., Chemat, F., 2012. Lab and pilot-scale ultrasound-assisted water extraction of polyphenols from apple pomace. Journal of Food Engineering 111 (1), 73–81.

Plaza, L., Sánchez-Moreno, C., Elez-Martínez, P., de Ancos, B., Martín-Belloso, O., Cano, M.P., 2006. Effect of refrigerated storage on vitamin C and antioxidant activity of orange juice processed by high-pressure or pulsed electric fields with regard to low pasteurization. European Food Research and Technology 223 (4), 487–493.

Poojary, M.M., Dellarosa, N., Roohinejad, S., Koubaa, M., Tylewicz, U., Gómez-Galindo, F., Saraiva, J.A., Dalla Rosa, M., Barba, F.J., 2017. Influence of innovative processing on γ-Aminobutyric acid (GABA) contents in plant food materials. Comprehensive Reviews in Food Science and Food Safety 16 (5), 895–905.

Ramazzina, I., Berardinelli, A., Rizzi, F., Tappi, S., Ragni, L., Sacchetti, G., Rocculi, P., 2015. Effect of cold plasma treatment on physico-chemical parameters and antioxidant activity of minimally processed kiwifruit. Postharvest Biology and Technology 107, 55–65.

Ramazzina, I., Tappi, S., Rocculi, P., Sacchetti, G., Berardinelli, A., Marseglia, A., Rizzi, F., 2016. Effect of cold plasma treatment on the functional properties of fresh-cut apples. Journal of Agricultural and Food Chemistry 64 (42), 8010–8018.

Roseiro, L.B., Duarte, L.C., Oliveira, D.L., Roque, R., Bernardo-Gil, M.G., Martins, A.I., Rauter, A.P., 2013. Supercritical, ultrasound and conventional extracts from carob (*Ceratonia siliqua* L.) biomass: effect on the phenolic profile and antiproliferative activity. Industrial Crops and Products 47, 132–138.

Sánchez-Moreno, C., Plaza, L., Elez-Martínez, P., De Ancos, B., Martín-Belloso, O., Cano, M.P., 2005. Impact of high pressure and pulsed electric fields on bioactive compounds and antioxidant activity of orange juice in comparison with traditional thermal processing. Journal of Agricultural and Food Chemistry 53 (11), 4403–4409.

Sánchez-Moreno, C., Plaza, L., De Ancos, B., Cano, M.P., 2006. Impact of high-pressure and traditional thermal processing of tomato purée on carotenoids, vitamin C and antioxidant activity. Journal of the Science of Food and Agriculture 86 (2), 171–179.

Sarangapani, C., O'Toole, G., Cullen, P., Bourke, P., 2017. Atmospheric cold plasma dissipation efficiency of agrochemicals on blueberries. Innovative Food Science & Emerging Technologies 44, 235–241.

Saucedo-Reyes, D., Marco-Celdrán, A., Pina-Pérez, M.C., Rodrigo, D., Martìnez-López, A., 2009. Modeling survival of high hydrostatic pressure treated stationary- and exponential-phase *Listeria innocua* cells. Innovative Food Science & Emerging Technologies 10, 135–141.

Surowsky, B., Schlüter, O., Knorr, D., 2014. Interactions of non-thermal atmospheric pressure plasma with solid and liquid food systems: a review. Food Engineering Reviews 7 (2), 82–108.

Tappi, S., Ramazzina, I., Rizzi, F., Sacchetti, G., Ragni, L., Rocculi, P., 2018. Effect of plasma exposure time on the polyphenolic profile and antioxidant activity of fresh-cut apples. Applied Sciences 8 (10), 1939.

Teissie, J., Golzio, M., Rols, M.P., 2005. Mechanisms of cell membrane electropermeabilization: a minireview of our present (lack of?) knowledge. Biochimica et Biophysica Acta 1724 (3), 270–280.

Tian, Y., Xu, Z., Zheng, B., Lo, Y.M., 2013. Optimization of ultrasonic-assisted extraction of pomegranate (*Punica granatum* L.) seed oil. Ultrasonics Sonochemistry 20 (1), 202–208.

Tiwari, B.K., 2015. Ultrasound: a clean, green extraction technology. TRAC Trends in Analytical Chemistry 71, 100–109.

Toepfl, S., Mathys, A., Heinz, V., Knorr, D., 2006. Potential of high hydrostatic pressure and pulsed electric fields for energy efficient and environmentally friendly food processing. Food Reviews International 22 (4), 405–423.

Tylewicz, U., Aganovic, K., Vannini, M., Toepfl, S., Bortolotti, V., Dalla Rosa, M., Oey, I., Heinz, V., 2016. Effect of pulsed electric field treatment on water distribution of freeze-dried apple tissue evaluated with DSC and TD-NMR techniques. Innovative Food Science & Emerging Technologies 37, 352–358.

Tylewicz, U., Tappi, S., Mannozzi, C., Romani, S., Dellarosa, N., Laghi, L., Ragni, L., Rocculi, P., Dalla Rosa, M., 2017. Effect of pulsed electric field (PEF) pre-treatment coupled with osmotic dehydration on physico-chemical characteristics of organic strawberries. Journal of Food Engineering 213, 2–9.

USFDA CFSAN-FDA (Center for Food Safety and Applied Nutrition), 2000. Kinetics of Microbial Inactivation for Alternative Food Processing Technologies. U.S. Food and Drug Administration, Washington.

Valdez-Fragoso, A., Mújica-Paz, H., Welti-Chanes, J., Torres, J.A., 2011. Reaction kinetics at high pressure and temperature: effects on milk flavor volatiles and on chemical compounds with nutritional and safety importance in several foods. Food and Bioprocess Technology 4, 986–995.

Virot, M., Tomao, V., Le Bourvellec, C., Renard, C.M., Chemat, F., 2010. Towards the industrial production of antioxidants from food processing by-products with ultrasound-assisted extraction. Ultrasonics Sonochemistry 17 (6), 1066–1074.

Weaver, J.C., Chizmadzhev, Y.A., 1996. Theory of electroporation: a review. Bioelectrochemistry and Bioenergetics 41, 135–160.

Wiktor, A., Sledz, M., Nowacka, M., Rybak, K., Chudoba, T., Lojkowski, W., Witrowa-Rajchert, D., 2015. The impact of pulsed electric field treatment on selected bioactive compound content and color of plant tissue. Innovative Food Science & Emerging Technologies 30, 69–78.

Wiktor, A., Sledz, M., Nowacka, M., Rybak, K., Witrowa-Rajchert, D., 2016. The influence of immersion and contact ultrasound treatment on selected properties of the apple tissue. Applied Acoustics 103 (Part B), 136–142.

Witrowa-Rajchert, D., Wiktor, A., Sledz, M., Nowacka, M., 2014. Selected emerging technologies to enhance the drying process: a review. Drying Technology 32 (11), 1386–1396.

Xu, L., Garner, A.L., Tao, B., Keener, K.M., 2017. Microbial inactivation and quality changes in orange juice treated by high voltage atmospheric cold plasma. Food and Bioprocess Technology 10, 1–14.

Zheng, L., Sun, D.W., 2006. Innovative applications of power ultrasound during food freezing processes—a review. Trends in Food Science & Technology 17, 16–23.

CHAPTER

9 Use of functional microbial starters and probiotics to improve functional compound availability in fermented dairy products and beverages

F. Patrignani, PhD [1,2], M. D'Alessandro[1], L. Vannini[1,2], R. Lanciotti[1,2]

[1]*Department of Agricultural and Food Sciences, Campus of Food Science, University of Bologna, Cesena, Italy;* [2]*CIRI - Agrifood (Interdepartmental Centre of Industrial Agrifood Research), University of Bologna, Cesena, Italy*

1. Introduction

The spontaneous fermentation of food has been used by mankind for more than 5000 years. It involved different products such as alcoholic beverages, dairy and meat products such as salami, and vegetable ones such as sauerkraut, olives, or Asian vegetables. It was only about 100 years ago that microbiologists recognized fermentation to be the result of useful microbial activity, and thus, individual microorganisms suitable for the fermentation processes were identified. In general, lactic acid bacteria (LAB) from different genera, such as *Lactobacillus*, *Streptococcus*, and *Leuconostoc,* are used in fermented foods, but also yeasts and molds can contribute to food fermentations. Basically, fermentative processes are guided by the microorganisms that act as a starter and distinguished by the primary metabolites produced (Gille et al., 2018): alcohol and carbon dioxide (principally yeast), acetic acid (*Acetobacter*), lactic acid (LAB), propionic acid (*Propionibacterium*), and ammonia and fatty acids (*Bacillus*, molds).

The most employed microorganisms for fermented food production are reported in Table 9.1.

Starter cultures are recognized as GRAS (generally recognized as safe) microorganisms due to their long history of use in food production. However, recently, the European Food Safety Agency (EFSA) established that microorganisms used as starter cultures are expected to meet several requirements and to satisfy the qualified presumption of safety (QPS) (EFSA, 2019).

The importance of fermented foods in human diet is well established even if their role has been evolved in the time. Historically, fermentation has been employed to

Sustainability of the Food System. https://doi.org/10.1016/B978-0-12-818293-2.00009-4

Table 9.1 Microorganisms generally involved in food fermentation.

Food	Species
Parmesan	*Lactobacillus helveticus, Lactobacillus delbrueckii* subsp. *bulgaricus, Streptococcus thermophilus*
Cheddar/Caciotta cheese	*Lactococcus lactis* subsp. *lactis/cremoris*
Squacquerone/ Stracchino like cheese	*S. thermophiles*
Sauerkraut	*Lactococcus mesenteroides, L. lactis*
Lupins, lentils, peas	*Lactobacillus acidophilus, Lactobacillus buchneri, Lactobacillus fermenti, Lactobacillus plantarum*
Fermented cassava	*L. plantarum*
Fermented fruit juice	*Lactobacillus casei*
Fermented vegetable juice	*L. plantarum, L. casei, Lactococcus lactis, Lactobacillus xylosus, Lactobacillus bavaricus*
Tempeh	*Rhizopus oligosporus*
Fermented soy milk	*Streptococcus salivarius ssp. thermophilus, L. delbrueckii* subsp. *bulgaricus*
Soy sauce	*Aspergillus oryzae, Pediococcus* species, *Weissella, Lactobacillus*
Vinegar	*Acetobacter* spp.

increase safety, shelf-life, texture, and sensory properties of foods (Rezac et al., 2018) since it acts as a low-cost method inhibiting spoilage and/or foodborne pathogenic microorganisms. Food preservation activities mainly result from acidification of the matrix, competition for nutrients, and the production of antimicrobial compounds such as bacteriocins (Siroli et al., 2015). Only recently, the use of fermented foods has arisen as important dietary strategy to improve human well-being due to the functional features that starter or adjunct cultures can impart to the food product. On the other side, fermented foods can often be used as carriers for probiotic bacteria, especially dairy products (Patrignani et al., 2016), whose economic impact is significant in food industry (Hatti-Kaul et al., 2018). On the other hand, during food fermentation, microorganisms may release in situ vitamins, bioactive and antioxidant compounds, and unsaturated fatty acids or contribute to inactivate antinutritional factors such as phytates (Rezac et al., 2018). Successful fermentation mainly depends on the microbial biodiversity and the microbial starter cultures used to induce an appreciated end product. The use of selected and tailor-made starter cultures is a common means of accelerating and guiding the fermentation process, thus improving the food quality and functionality. Also, the use of selected starter cultures isolated from the naturally occurring microbiota of fermented foods can represent a strategy to guarantee a tailor-made culture for each considered production assuring also specified or increased functionalities. So, a fermented product represents by itself an example of functional foods. On the other side, these represent one of the fastest growing areas in the global food industry because these considered a dietary strategy to reduce the incidence of illness in the humankind. Generally, probiotic bacteria are defined as "*live microorganisms which when consumed in adequate numbers confer a health benefit on the host*." Scientific evidences indicate that ingestion of probiotic cultures exerts health benefits both at gastrointestinal level and in the respiratory and urogenital tracts. Conditions such as lactose intolerance, diarrhea, gastroenteritis, irritable bowel syndrome, inflammatory bowel disease, depressed immune function, cancer, and genitourinary tract infections have all been reported to benefit from probiotics. In any case, using microorganisms like both starter cultures and probiotic adjuncts, these approaches must be based on specific food biotechnological processes, to achieve specific food functionalities. Although genetic engineering is practiced for this aim, the use of well-selected and characterized wild and robust strains can offer a further competitive advantage to obtain foods with new nutritional value and functional and organoleptic properties. Moreover, the increasing demand for high-value health-promoting foods has encouraged the industries and, consequently, the research field, to investigate new emerging food processes able to guarantee safe functional products (Burns et al., 2015). Currently, the most important and interesting applications have been studied and applied for the dairy ingredients or products due to their great potential as functional and nutraceuticals.

However, the criteria used to assess the suitability of a strain to be used in food product as a starter or functional adjunct are, generally, the kinetics and total production of acids, pH variation, loss of substances with high nutritional value, and decrease in nitrates and in biogenic amines (Karovicová & Kohajdová, 2005). Moreover, the ability of the microbial culture to grow on the substrate, the type of metabolism, and its ability to impart desirable sensory properties represent also selection criteria.

Moreover, the challenge of this era is to sustainably increase the global food production system, including also that of functional foods and probiotic products. This challenge can be achieved throughout several solutions starting from the use of food industrial by-products as raw material to use for fermentation but also using optimized starter/costarter cultures and tailor-made biotechnological processes (Iriondo-Dehond et al., 2018). Moreover, the proper choice of the initial by-product can be useful to define the final features of the final fermented functional food. For example, by-products from dairy, meat, fish, and seafood contain high amounts of protein that can be transformed, in their turn, due to hydrolytic microbial activities, in functional peptides or bioactive ingredients with functional role. In addition, also by-products from fruit material and vegetables were used for the formulation of health-promoting dairy products.

Thus, in this scenario, this chapter is aimed to give an overview on the applicative potentialities of starter microorganisms and probiotics to increase in sustainable way quality, safety, and functionality of fermented dairy products and fruit and vegetable beverage contributing to enhance human well-being.

2. Fermented dairy products

Fermented dairy products are economically the most important food sector with an estimated value of over 80 billion euros. Opportunities for innovations and developments in fermented milks are also represented by added-value products such as those fortified with bioactive ingredients, including fibers, phytosterols, whey-based ingredients, antioxidant vitamins, and isoflavones that provide specific health benefits beyond basic nutrition (Hatti-Kaul et al., 2018). From 2000 to 2018, 50 eligible studies have pointed out the development of innovative and health-promoting dairy products using sustainable ingredients, 88% of which used side streams from plant materials. In particular, fruit and vegetable by-products (citrus and tomato side-streams) were used as ingredients in dairy formulations (FAO, 2017). Also, the food by-product application in dairy sector can be divided into two parts having the purpose of (1) improving shelf life, safety, stability, and sensory quality of the final product or (2) enhancing its health-promoting effects (Iriondo-Dehond et al., 2018).

Among the over 3500 traditional fermented foods existing worldwide, fermented milks have long been an important component of nutrition and diet. The species used in fermentation process (*Lactobacillus bulgaricus*, *Streptococcus thermophilus*) or added as probiotics (*Lactobacillus acidophilus, Bifidobacterium bifidum, Bifidobacterium lactis, Lactobacillus casei*, and *Lactobacillus rhamnosus*) are reported to improve digestion and bioavailability of milk constituents, to inhibit harmful bacteria of gastrointestinal tract, and to alleviate lactose intolerance and hypocholesterolemic effect. During fermentation and the following storage, several bioactive peptides can also be released by proteolytic cultures, which are endowed with extracellular proteinases and some intracellular ones such as dipeptidases, tripeptidases, aminopeptidases, and endopeptidases.

The antihypertensive peptides Val-Pro-Pro and Ile-Pro-Pro, which inhibit angiotensin-converting enzyme, have been isolated from various types of dairy products, including processed milk, cheese, sour milk, milk fermented by *Lactobacillus helveticus* (FitzGerald et al., 2004), while a peptide active against hypertension derived from β-casein (Ser-Lys-Val-Tyr-Pro-Phe-Pro-Gly Pro-Ile) has been detected in milks produced with *Lactobacillus delbrueckii* ssp. *bulgaricus species*. However, it should be considered that a wide range of bioactive compounds can be produced in cheeses since bacterial proteolytic activities play a key role also during their ripening.

Several studies evidenced the potential of fermented dairy products to protect from cancer. Larsson et al. (2008), in a study among over 80,000 Swedish volunteers, highlighted the decrease of risk for bladder in patients consuming sour milk and yogurt. Also, a reduced tumor growth rate and increased lymphocyte proliferation were observed in a mouse model of breast cancer following treatment with *L. acidophilus* isolated from home-made traditional yogurt (Maroof et al., 2012).

A strategy to produce a novel functional yogurt with different biological activities has been evaluated by El-Fattah et al. (2018) who fortified yogurts with whey protein concentrate (2% w/v) and skim milk powder (2% w/v) fermented by yogurt cultures and highly proteolytic cultures of *Lactococcus cremoris* and

Lactococcus lactis, *L. helveticus* Lh-B 02, and *L. rhamnosus* B-1445. Exploiting the diverse proteinases and peptidases system of the mixed starters, a wide range of peptides with inhibitory effect on hypertensive (anti-ACE), thrombin, and cholesterol micellar solubility were produced, improving the biological activities of the product. Such properties were retained or even improved during a 14 days' cold storage. Moreover, the selected mixed cultures showed a relevant role in increasing the γ-aminobutyric acid (GABA) content (up to 1.64 mg/100 mL) through the action of the glutamate decarboxylase, which converts glutamate into the amino acid. It is well known that GABA has physiological functions through important processes such as neurotransmission and antihypertensive activities (Hayakawa et al., 2004).

The scope for health promotion and/or support by probiotics such as LAB is, however, in great expansion beyond the improving of intestinal health, also to include additional benefits such as antiobesity, antidiabetic, and cholesterol-lowering effects. Park et al. (2018) investigated the cholesterol lowering and microbiota modulatory potential of *L. rhamnosus* BFE5264, isolated from Maasai fermented milk, by using a mouse model receiving a high-cholesterol diet. Authors observed a significant reduction of the serum cholesterol level following *L. rhamnosus* BFE5264 administration, changes of gut microbiota, and accumulation of short chain fatty acids. The selected strain also beneficially influenced the regulation of cholesterol metabolism in the liver in a similar way to that resulting from statin treatment, a drug-inhibiting cholesterol biosynthesis in the liver. Patrignani et al. (2016) used the same strain *L. rhamnosus* BFE5264, as a costarter, to produce a functional fermented milk using high-pressure homogenized milk (60 MPa) verifying also the organoleptic, textural, and functional features of the obtained product.

Supplementation of dairy products with probiotic strains such as *Bifidobacterium* and *L. acidophilus*, claimed to have numerous health benefits (Butel, 2014), has become the focus of an increasing number of researches and also an important issue for industry and regulating authorities alike. However, probiotics, before colonizing the human epithelium, need to survive to several stresses including technological ones during the product process and human barriers. Among the possible strategies to protect probiotics, the microencapsulation, using proper matrix able to convey them limiting chemicophysical and environmental stresses, has been successfully applied by several authors. Patrignani et al. (2017) produced sodium alginate/vegetable oil microcapsules with entrapped *Lactobacillus paracasei* A 13 and *Lactobacillus salivarius* subsp. *salivarius* CET 4063 cells to produce functional fermented milks. This strategy proved to be effective in maintaining high probiotic viability over 1 month of refrigerated storage also through a limited and slower product acidification, increasing probiotic resistance to the simulated digestion processes (Fig. 9.1).

Indeed, Patrignani et al. (2019a) successfully produced probiotic cheese, i.e., Squacquerone, as a carrier of *Lactobacillus crispatus* BC4, a vaginal strain characterized by both antimicrobial activity against urogenital pathogens and foodborne microorganisms and interesting technological properties. The probiotic strain used as adjunct culture maintained high viability values (7 Log CFU/g) for all the refrigerated storage and also during a simulated stomach-duodenum passage. Moreover, it

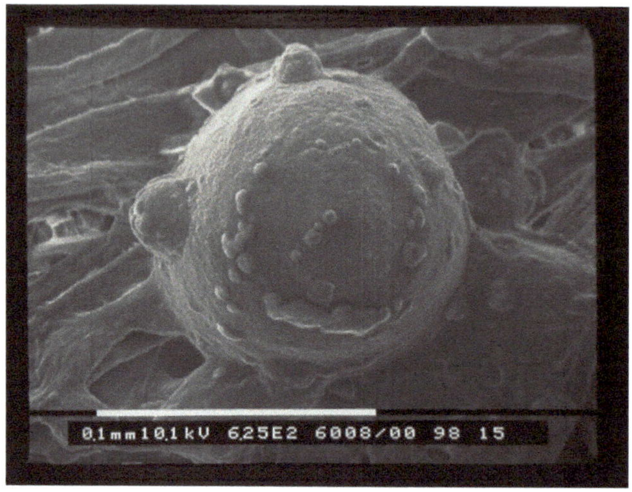

FIGURE 9.1

Example of incapsulated *Lactobacillus paracasei* A13 adopting the process condition described by Patrignani et al. (2017).

positively affected the organoleptic features of the produced cheese. On the other hand, the ability to impart good sensory properties to the final product should be included in the selection criteria of the probiotic culture since the overall properties of the products are important product traits affecting consumers' acceptability. Also, the use of nonthermal technologies for milk treatment, instead of traditional pasteurization, can be a good strategy to increase probiotic viability. Patrignani et al. (2009) used high-pressure homogenization (HPH) treatment at 60 MPa to maintain high level of *L. acidophilus* 08 and *L. paracasei* A13 inoculated in whole milk for the production of fermented milks. HPH treatments performed at 50 MPa have been reported to improve also *L. paracasei* A13 functional properties such as hydrophobicity and resistance to simulated gastric digestion (Tabanelli et al., 2013), and IgA response of mice small intestine in a strain- and feeding period-dependent way (Tabanelli et al., 2012). Such enhancement of strain functionality is related to modification of the outer cellular structures induced by the HPH treatment (Tabanelli et al., 2015) and has been exploited for the production of probiotic Caciotta cheese containing HPH-treated (50 MPa) *L. paracasei* A 13 cells as adjunct cultures (Burns et al., 2015). Results confirmed that viability of *L. paracasei* A13 was unaffected by the sublethal HPH treatment maintaining high strain viability in cheese up to 21 days of Caciotta ripening. Although the HPH treatment negatively affected the IL-10 secretion, the IgA production was not influenced by the treatment.

3. Fermented functional drinks and beverage

Fermentation is considered a useful process to preserve also plant materials without detrimental effect on nutritional value. The fermentation of vegetable and fruit can be affected by various microorganisms in relation to the origin of the raw material.

Generally, LAB and yeasts are used in European and American countries, whereas in Asiatic regions, a large number of products (especially vegetables) are fermented by molds. However, the most extensively used process involves LAB. Fermentation of fruit juices by the use of probiotic bacteria or protective cultures is becoming a very interesting topic to overcome the loss of nutritional value of pasteurized/sterilized juices. Differently, the fermentation of raw vegetable material is a usual practice both in developed and in developing countries. As previously reported for fermented milk, several vegetable by-products can be used for the formulation of new functional vegetable drinks. For example, bananas, plantains, and pineapples have some of the highest loss rates among fruits and vegetables during storage and transportation. However, fermentation is very important to increase sensory end texture properties, destroy natural occurring toxins, improve digestibility, and enrich fermented foods of desired metabolites such as lactic acid, amino acids, and phenolic compounds, enriching their nutritional values. When fermentation takes place, complex microbial interactions occur. In general, the quality of the raw material affects significantly the quality and safety of the final fermented product. The most used commercial strains involved in fruit juice fermentation belong to the species of *L. casei*, *Lactobacillus plantarum*, *Lactobacillus xylosus*, *Lactobacillus brevis*, *Lactobacillus reuteri*, *B. lactis*, and *B. bifidum* (Siroli et al., 2019), whereas strains of *Leuconostoc mesenteroides* are applied to sauerkraut fermentation. *L. acidophilus*, *Lactobacillus buchneri*, and *Lactobacillus fermenti* are employed in lupins, lentils, and other legumes, whereas *L. plantarum* is used for Cassava fermentation. *Aspergillus oryzae*, several lactobacilli, *Pediococcus* species, and yeasts are for the fermentation of soy products.

The market of fruit juices and beverages is actually indicating a dynamic development worldwide. In particular, functional vegetable beverages with probiotic bacteria are considered as an attractive option for people not consuming dairy products. Moreover, fruit juices have been considered as novel suitable carrier for the correct intake of probiotic bacteria, as they have vitamins, minerals, and antioxidant compounds providing an appropriate growth substrate for microorganisms and a strong health appeal for consumers. Fermentation of fruit juice by probiotic bacteria can increase viability of the cells and improve juice functional aspects (Mantzourani et al., 2019). Several current researches have recently discussed on fermented fruit juice production using various probiotic strains (Betoret et al., 2017; Mantzourani et al., 2018), highlighting the enriching benefits of the obtained juices after the fermentation process. Mantzourani et al. (2019) used *L. plantarum* ATCC 14917, gifted with good technological features, to produce a novel fermented pomegranate beverage. The use of this strain led to the production of a functional fruit beverage with low alcoholic degree having higher levels of total phenolic content and antioxidant activity. These effects were due to the lactic acid fermentation operated by the strain and also due to the strain production of β-galactosidase catalyzing the release of phenolic compounds from the bonded sugars, resulting in an increase in the antioxidant activity after fermentation (Sharma et al., 2018). *L. plantarum* ATCC 14917 maintained high viability (above 8.8 log cfu/mL) complying the FIL-IDF criteria for probiotic strains.

Betoret et al. (2017) enriched mandarin juice with *L. salivarius* spp. *salivarius* CET 4063, with anti-*Helicobacter pylori* effect. The authors, to support the strain availability in juice, added different concentrations of trehalose and treated the juice with several HPH treatments demonstrating that both food matrix and processing affected functional properties of CET 4063. The presence of trehalose improved cell hydrophobicity and influenced the effects promoted by HPH, guarantying also a further protection on further applied stress conditions. A further step of this research was carried out by Calabuig-Jiménez et al. (2018) who proposed the microencapsulation of the same strain by HPH to protect it in mandarin juice. The microencapsulation by homogenization at 70 MPa with alginate was a promising strategy to protect *L. salivarius* spp. *salivarius* during gastrointestinal digestion process and storage. Mantzourani et al. (2018) developed a novel functional cherry juice using *L. plantarum* ATCC 14917 immobilized in a delignified wheat bran carrier. The fermentation, performed at 30°C for 24 h, gave rise to a fermented cherry juice endowed with higher total phenolic content (214–264 mg GAE/100 mL) compared with fermented juice with free cells (165–199 mg GAE/100 mL) and nonfermented juice (135–169 mg GAE/100 mL). Immobilized cells retained their viability in higher levels (9.95 log cfu/mL at the fourth week) compared with free cells (7.36 log cfu/mL at the fourth week) without detrimental effects on juice sensory properties. Amorim et al. (2018) proposed *Meyerozyma caribbica*, isolated from pineapple, to obtain a fermented pineapple-based beverage, using, as control, the juice obtained from the commercial probiotic *Saccharomyces cerevisiae* var. *boulardii*. The juice produced with *M. caribbica* was gifted, with respect the control, of higher concentrations of residual glucose and fructose, lower concentration of acetic acid, higher total phenolic compounds, catechin, chlorogenic acid, vanillin, and ferulic acid. *M. caribbica*, during juice-refrigerated storage, maintained cell loads higher than 7.0 log CFU/mL over 21 days. Also, Kanwar et al. (2016) studied the variability of the gene ATF, responsible for ester synthesis during fermentation, in 10 *S. cerevisiae* strains, isolated from fermented foods of Western Himalayas. The results highlighted several significant variations in ATF1 gene sequences, suggesting differences in aroma and flavor of the brewing products obtained using the different strains.

Also, the development of fermented functional vegetable drinks and beverage is very appealing. In fact, vegetable beverages are perceived by consumers as "healthy" foods because of their low content of sodium, cholesterol, and fat and because they are rich in vitamin C, polyphenols, and flavonoids that contribute to the good antioxidant properties of these products (Patrignani et al., 2009a,b). Similarly, the use of vegetable by-products could be an important strategy to increase the process sustainability and to increase the circular economy model. For these reasons, their market is increased in recent years (Kregiel, 2015). Although almond rice-milk is well known, soymilk and soy products are certainly the most requested and widespread among western consumers (Riciputi et al., 2016). Soymilk is an easily digestible food, and it is generally characterized by high protein content, a moderate ratio of β-conglycinin and glycinin and high sugar content (Kaneko et al., 2011). Protein and fatty components in soybeans are high in containing all the essential amino acids in amounts that closely match those required by humans. Also, soybean lipids contain a high proportion of unsaturated fatty acids, including oleic, linoleic, linolenic acids, and minor lipid components as phospholipids, phytosterols, and tocopherols with recognized health-promoting features (Medic et al., 2014). In addition, soy-based foods have received increased attention in health conscious societies, especially in Western countries, due to their reported beneficial effects on menopausal symptoms and metabolic-related diseases (Cao et al., 2018). Also the consumption of carrot juice, the most popular nonalcoholic beverages consumed in Germany and northern European countries (Sharma et al., 2012), is increased in recent years (Sharma et al., 2012). In fact, it is an important natural source of antioxidants, such as α- and β-carotene, and precursors of vitamin A and polyacetylene that have beneficial health and antitumor activities (Sharma et al., 2012; Zhang et al., 2016). Also, its antianemic activity and benefits in the healing of wounds are documented (Nadeem et al., 2018). However, carrot and soy beverages due to their high pH and the high content in sugars can favor the growth of spoilage and pathogenic microorganisms (Patrignani et al., 2009a,b). Recent outbreaks of foodborne diseases have also been attributed to unpasteurized juices having lower pH values such as orange and apple juices contaminated with pathogenic agents such as *Salmonella* spp., *Escherichia coli* O157: H7, and *Listeria monocytogenes* (Callejón et al., 2015). Recently, the need to increase shelf life of such products, without detrimental effect on sensory, functional, and nutritional properties, has generated a growing interest toward mild heat treatments in combination with refrigeration or nonthermal biotechnologies including HHP, HPH (Patrignani et al., 2019b), and lactic fermentation of vegetable matrices (Tremarin et al., 2017). Lactic acid fermentation is considered an important technology due to the increasing quantity of raw materials processed in this way in the food industry (Di Cagno et al., 2013). The main reasons for this interest are the nutritional, physiological, and hygienic aspects of the process and the related costs of implementation and production (Karovicová and Kohajdová, 2005). Literature reports an increase in fermented vegetable juices produced mainly from cabbage, red beets, carrots, celery, and tomatoes (Di Cagno et al., 2013). It is reported that fermentation has been used to increase the bioavailability of vitamins, minerals, and isoflavones in soy, as well as to modify its flavor, improve its stability, and even create new food products (Riciputi et al., 2016). Recently, Siroli et al. (2019) proposed nisin producer *L. lactis* strains as promising biopreservatives to obtain more safe, controlled, and reproducible fermentations of soymilk and carrot beverage (Siroli et al., 2016; Ho et al., 2018). Nisin was the first bacteriocin characterized and allowed as a food preservative in the European Union. In fact, nisin, particularly the variant Z, has a high solubility and stability in different food systems with a wide antimicrobial spectrum, effective against gram-positive bacteria (Gharsallaoui et al., 2016). However, few studies about in situ production of nisin are available (de Oliveira Junior et al., 2015). Siroli et al. (2019) demonstrated the potential of three nisin-producing *L. lactis* strains for microbiological stabilization of fermented carrot juice and soymilk.

Chwastek (2015) used a similar approach to optimize an innovative process for the production of a novel probiotic beverage produced from red beet juice and supplemented with waste sucrose syrup. The presence of the functional strains increased in the juice the level of total phenolic compounds.

Vanajakshi et al. (2015) proposed the optimization of a probiotic beverage based on Moringa leaves and beetroot, known for their nutritional and health benefits for malnourished people all over the world. The fermentation was guided by *L. plantarum* and *Enterococcus hirae* for 48 h at 37°C. The resulting fermented beverage showed a strong antibacterial activity against foodborne pathogens. It also exhibited radical scavenging activity (20.79%) with a phenolic content of 5 mg/mL and had minerals (mg/mL) like 11.8 of calcium and 0.2 of iron.

Panghal et al. (2018) highlighted also the application of cereals in the development of nondairy probiotic drink based on the use of multigrains. In general, cereal-based beverages are rich in phenolic compounds, sterols, vitamins B and E, minerals, and fibers (Fernandes et al., 2019). However, they can contain phytate acting as antinutritional factors by forming insoluble complexes with cations at physiological pH values and inhibiting mineral absorption. For this, this feature is considered in the selection of strains for the production of functional/fermented vegetable. To improve the mineral bioavailability, several approaches were proposed considering yeasts or/and LAB naturally present in flour or added as starter cultures used in mono- or coculture (Alvarez-Martin et al., 2008).

4. Conclusion

Microorganisms can be considered an important biotechnological strategy to improve in sustainable way food product functionalities, increasing also their overall quality and safety. Many traditional naturally fermented products can be considered as natural source for the isolation, characterization, and exploitation of microorganisms of industrial interest maintaining, at the same time, a strong connection with tradition and local culture. It is, thus, mandatory to select and optimize the performances of tailor-made cultures to obtain innovative products with specific features and satisfying a new generation of consumers. The optimization of the fermentation processes is mandatory also to increase the whole process sustainability as well as the use of sustainable nonthermal technologies can be considered a useful tool to increase specific microorganism functionalities. The exploitation of food by-products as source of sustainable innovative ingredients can be desirable for the entire waste management system and for the development of innovative and healthy sustainable foods. In addition, this approach can be considered a tool to diversify the production of functional foods, creating the possibility of satisfying a market niche based on functional and sustainable products. Although it will be necessary to understand the consumer perception and concerns on the use of by-products in food formulation, this can represent a great challenge for the scientists.

References

Alvarez-Martin, P., Florez, A.B., Hernández-Barranco, A., Mayo, B., 2008. Interaction between dairy yeasts and lactic acid bacteria strains during milk fermentation. Food Control 19 (1), 62–70.

Amorim, J.C., Piccoli, R.H., Duarte, W.F., 2018. Probiotic potential of yeasts isolated from pineapple and their use in the elaboration of potentially functional fermented beverages. Food Research International 107, 518–527.

Betoret, E., Calabuig-Jiménez, L.A.U.R.A., Patrignani, F., Lanciotti, R., Dalla Rosa, M., 2017. Effect of high pressure processing and trehalose addition on functional properties of Mandarin juice enriched with probiotic microorganisms. LWT-Food Science and Technology 85, 418–422.

Burns, P.G., Patrignani, F., Tabanelli, G., Vinderola, G.C., Siroli, L., Reinheimer, J.A., Gardfini, F., Lanciotti, R., 2015. Potential of high pressure homogenisation on probiotic Caciotta cheese quality and functionality. Journal of functional foods 13, 126–136.

Butela, M.J., 2014. Probiotics, gut microbiota and health. Medecine et Maladies Infectieuses 44 (1), 1–8.

Calabuig-Jiménez, L., Betoret, E., Betoret, N., Patrignani, F., Barrera, C., Seguí, L., Dalla Rosa, M., 2018. High pressures homogenization (HPH) to microencapsulate *L. salivarius* spp. *salivarius* in Mandarin juice. Probiotic survival and in vitro digestion. Journal of Food Engineering.

Callejón, R.M., Rodríguez-Naranjo, M.I., Ubeda, C., Hornedo-Ortega, R., Garcia-Parrilla, M.C., Troncoso, A.M., 2015. Reported foodborne outbreaks due to fresh produce in the United States and European Union: trends and causes. Foodborne Pathogens and Disease 12 (1), 32–38.

Cao, Z.H., Green-Johnson, J.M., Buckley, N.D., Lin, Q.Y., 2018. Bioactivity of soy-based fermented foods: a review. Biotechnology Advances.

Chwastek, A., Klewicka, E., Klewicki, R., Sójka, M., 2016. Lactic acid fermentation of red beet juice supplemented with waste highbush blueberry–sucrose osmotic syrup as a method of probiotic beverage production. Journal of Food Processing and Preservation 40 (4), 780–789.

de Oliveira Junior, A.A., de Araújo Couto, H.G.S., Barbosa, A.A.T., Carnelossi, M.A.G., de Moura, T.R., 2015. Stability, antimicrobial activity, and effect of nisin on the physico-chemical properties of fruit juices. International Journal of Food Microbiology 211, 38–43.

Di Cagno, R., Coda, R., De Angelis, M., Gobbetti, M., 2013. Exploitation of vegetables and fruits through lactic acid fermentation. Food Microbiology 33 (1), 1–10.

EFSA (2019). https://www.efsa.europa.eu/en/topics/topic/qualified-presumption-safety-qps.

El-Fattah, A.A., Sakr, S., El-Dieb, S., Elkashef, H., 2018. Developing functional yogurt rich in bioactive peptides and gamma-aminobutyric acid related to cardiovascular health. LWT – Food Science and Technology 98, 390–397. https://doi.org/10.1016/j.lwt.2018.09.022.

FAO FAOSTAT. Available online: http://www.fao.org/faostat (accessed on 18 December 2017).

Fernandes, C.G., Sonawane, S.K., SS, A., 2019. Cereal based functional beverages: a review. Journal of Microbiology, Biotechnology and Food Sciences 2019, 914–919. https://doi.org/10.15414/jmbfs.2018-19.8.3.914-919.

FitzGerald, R.J., Murray, B.A., Walsh, D.J., 2004. Hypotensive peptides from milk proteins. Journal of Nutrition 134 (4), 980S–988S. https://doi.org/10.1093/jn/134.4.980S.

Gharsallaoui, A., Oulahal, N., Joly, C., Degraeve, P., 2016. Nisin as a food preservative: part 1: physicochemical properties, antimicrobial activity, and main uses. Critical Reviews in Food Science and Nutrition 56 (8), 1262–1274. https://doi.org/10.1080/10408398.2013.763765.

Gille, D., Schmid, A., Walther, B., Vergères, G., 2018. Fermented food and non-communicable chronic diseases: a review. Nutrients 10 (4), 448. https://doi.org/10.3390/nu10040448.

Hatti-Kaul, R., Chen, L., Dishisha, T., Enshasy, H.E., 2018. Lactic acid bacteria: from starter cultures to producers of chemicals. FEMS Microbiology Letters 365 (20). https://doi.org/10.1093/femsle/fny213.

Hayakawa, K., Kimura, M., Kasaha, K., Matsumoto, K., Sansawa, H., Yamori, Y., 2004. Effect of a γ-aminobutyric acid-enriched dairy product on the blood pressure of spontaneously hypertensive and normotensive Wistar–Kyoto rats. British Journal of Nutrition 92 (3), 411–417. https://doi.org/10.1079/BJN20041221.

Ho, V.T.T., Lo, R., Bansal, N., Turner, M.S., 2018. Characterisation of *Lactococcus lactis* isolates from herbs, fruits and vegetables for use as biopreservatives against *Listeria monocytogenes* in cheese. Food Control 85, 472–483. https://doi.org/10.1093/femsle/fny213.

Iriondo-Dehond, M., Miguel, E., Del Castillo, M.D., 2018. Food byproducts as sustainable ingredients for innovative and healthy dairy foods. Nutrients 10, 1358. https://doi:10.3390/nu10101358.

Kaneko, S., Kumazawa, K., Nishimura, O., 2011. Studies on the key aroma compounds in soy milk made from three different soybean cultivars. Journal of Agricultural and Food Chemistry 59 (22), 12204–12209. https://doi.org/10.1021/jf202942h.

Kanwar, S.S., 2016. Fermentation of apple juice with a selected yeast strain isolated from the fermented foods of himalayan regions and its organoleptic properties. Frontiers in Microbiology 7, 1012. https://doi.org/10.3389/fmicb.2016.01012.

Kregiel, D., 2015. Health safety of soft drinks: contents, containers, and microorganism. BioMed Research International 2015. https://doi.org/10.1155/2015/128697.

Larsson, S.C., Andersson, S.O., Johansson, J.E., Wolk, A., 2008. Cultured milk, yogurt, and dairy intake in relation to bladder cancer risk in a prospective study of Swedish women and men. American Journal of Clinical Nutrition 88 (4), 1083–1087. https://doi.org/10.1093/ajcn/88.4.1083.

Mantzourani, I., Kazakos, S., Terpou, A., Alexopoulos, A., Bezirtzoglou, E., Bekatorou, A., Plessas, S., 2019. Potential of the probiotic *Lactobacillus plantarum* ATCC 14917 strain to produce functional fermented pomegranate juice. Foods 8 (1), 4. https://doi.org/10.3390/foods8010004.

Mantzourani, I., Nouska, C., Terpou, A., Alexopoulos, A., Bezirtzoglou, E., Panayiotidis, M., Plessas, S., 2018. Production of a novel functional fruit beverage consisting of cornelian cherry juice and probiotic bacteria. Antioxidants 7 (11), 163. https://doi.org/10.3390/antiox7110163.

Maroof, H., Hassan, Z.M., Mobarez, A.M., Mohamadabadi, M.A., 2012. *Lactobacillus acidophilus* could modulate the immune response against breast cancer in murine model. Journal of Clinical Immunology 32 (6), 1353–1359.

Medic, J., Atkinson, C., Hurburgh Jr., C.R., 2014. Current knowledge in soybean composition. Journal of the American Oil Chemists' Society 91 (3), 363–384.

Nadeem, M., Ubaid, N., Qureshi, T.M., Munir, M., Mehmood, A., 2018. Effect of ultrasound and chemical treatment on total phenol, flavonoids and antioxidant properties on carrot-grape juice blend during storage. Ultrasonics Sonochemistry 45, 1–6.

Panghal, A., Janghu, S., Virkar, K., Gat, Y., Kumar, V., Chhikara, N., 2018. Potential non-dairy probiotic products—a healthy approach. Food Bioscience 21, 80–89.

Park, S., Kang, J., Choi, S., Park, H., Hwang, E., Kang, Y., Ji, Y., 2018. Cholesterol-lowering effect of *Lactobacillus rhamnosus* BFE5264 and its influence on the gut microbiome and propionate level in a murine model. PLoS One 13 (8).

Patrignani, F., Burns, P., Serrazanetti, D., Vinderola, G., Reinheimer, J., Lanciotti, R., Guerzoni, M.E., 2009a. Suitability of high pressure-homogenized milk for the production of probiotic fermented milk containing *Lactobacillus paracasei* and *Lactobacillus acidophilus*. Journal of Dairy Research 76 (1), 74–82.

Patrignani, F., Serrazanetti, D.I., Mathara, J.M., Siroli, L., Gardini, F., Holzapfel, W.H., Lanciotti, R., 2016. Use of homogenisation pressure to improve quality and functionality of probiotic fermented milks containing *Lactobacillus rhamnosus* BFE 5264. International Journal of Dairy Technology 69 (2), 262–271.

Patrignani, F., Siroli, L., Parolin, C., Serrazanetti, D.I., Vitali, B., Lanciotti, R., 2019a. Use of *Lactobacillus crispatus* to produce a probiotic cheese as potential gender food for preventing gynaecological infections. PLoS One 14 (1). https://doi.org/10.1371/journal.pone.0208906.

Patrignani, F., Siroli, L., Serrazanetti, D.I., Braschi, G., Betoret, E., Reinheimer, J.A., Lanciotti, R., 2017. Microencapsulation of functional strains by high pressure homogenization for a potential use in fermented milk. Food Research International 97, 250–257.

Patrignani, F., Vannini, L., Kamdem, S.L.S., Lanciotti, R., Guerzoni, M.E., 2009b. Effect of high pressure homogenization on *Saccharomyces cerevisiae* inactivation and physico-chemical features in apricot and carrot juices. International Journal of Food Microbiology 136 (1), 26–31.

Patrignani, F., Mannozzi, C., Tappi, S., Tylewicz, U., Pasini, F., Castellone, V., Riciputi, Y., Rocculi, P., Romani, S., Caboni, M.F., Gardini, F., Lanciotti, R., Dalla Rosa, M., 2019b. (Ultra) High pressure homogenization potential on the shelf-life and functionality of kiwifruit Juice. Frontiers in Microbiology 10 article 246. https:/doi:10.3389/fmicb.2019.00246.

Rezac, S., Kok, C.R., Heermann, M., Hutkins, R., 2018. Fermented foods as a dietary source of live organisms. Frontiers in Microbiology 9. https://doi.org/10.3389/fmicb.2018.01785.

Riciputi, Y., Serrazanetti, D.I., Verardo, V., Vannini, L., Caboni, M.F., Lanciotti, R., 2016. Effect of fermentation on the content of bioactive compounds in tofu-type products. Journal of Functional Foods 27, 131–139.

Sharma, K.D., Karki, S., Thakur, N.S., Attri, S., 2012. Chemical composition, functional properties and processing of carrot, a review. Journal of Food Science & Technology 49 (1), 22–32.

Sharma, S., Kandasamy, S., Kavitake, D., Shetty, P.H., 2018. Probiotic characterization and antioxidant properties of *Weissella confusa* KR780676, isolated from an Indian fermented food. LWT - Food Science and Technology 97, 53–60.

Siroli, L., Camprini, L., Pisano, M.B., Patrignani, F., Lanciotti, R., 2019. Volatile molecule profiles and anti-*Listeria monocytogenes* activity of nisin producers *Lactococcus lactis* strains in vegetable drinks. Frontiers in Microbiology 10, 563.

Siroli, L., Patrignani, F., Serrazanetti, D.I., Tabanelli, G., Montanari, C., Gardini, F., Lanciotti, R., 2015. Lactic acid bacteria and natural antimicrobials to improve the safety and shelf-life of minimally processed sliced apples and lamb's lettuce. Food Microbiology 47, 74–84.

Siroli, L., Patrignani, F., Serrazanetti, D.I., Vannini, L., Salvetti, E., Torriani, S., Gardini, F., Lanciotti, R., 2016. Use of a nisin-producing *Lactococcus lactis* strain, combined with natural antimicrobials, to improve the safety and shelf-life of minimally processed sliced apples. Food Microbiology 54, 11–19.

Tabanelli, G., Burns, P., Patrignani, F., Gardini, F., Lanciotti, R., Reinheimer, J., Vinderola, G., 2012. Effect of a non-lethal High pressure homogenization treatment on the in vivo response of probiotic lactobacilli. Food Microbiology 32 (2), 302–307.

Tabanelli, G., Patrignani, F., Vinderola, G., Reinheimer, J.A., Gardini, F., Lanciotti, R., 2013. Effect of sub-lethal high pressure homogenization treatments on the in vitro functional and biological properties of lactic acid bacteria. LWT - Food Science and Technology 53 (2), 580–586.

Tabanelli, G., Vernocchi, P., Patrignani, F., Del Chierico, F., Putignani, L., Vinderola, G., Lanciotti, R., 2015. Effects of sub-lethal high-pressure homogenization treatment on the outermost cellular structures and the volatile-molecule profiles of two strains of probiotic lactobacilli. Frontiers in Microbiology 6, 1006.

Tremarin, A., Brandão, T.R., Silva, C.L., 2017. Inactivation kinetics of *Alicyclobacillus acidoterrestris* in apple juice submitted to ultraviolet radiation. Food Control 73, 18–23.

Vanajakshi, V., Vijayendra, S.V.N., Varadaraj, M.C., Venkateswaran, G., Agrawal, R., 2015. Optimization of a probiotic beverage based on Moringa leaves and beetroot. LWT-Food Science and Technology 63 (2), 1268–1273.

Zhang, Y., Liu, X., Wang, Y., Zhao, F., Sun, Z., Liao, X., 2016. Quality comparison of carrot juices processed by high-pressure processing and high-temperature short-time processing. Innovative Food Science & Emerging Technologies 33, 135–144.

CHAPTER

Globalization of technologies: pros and cons

10

Ester Betoret, PhD [1], Noelia Betoret, PhD [2]

[1] *Instituto de Agroquímica y Tecnología de Alimentos, Consejo Superior de Investigaciones Científicas, Paterna, Spain;* [2] *Universitat Politècnica de València, València, Spain*

1. Globalization, technologies, and food system

Globalization is the word used to describe the growing interdependence of the world's economies, cultures, and populations, brought about by cross-border trade in goods and services, technology, and flows of investment, people, and information (Peterson Institute for International Economics, 2019).

The driving force and impacts of globalization are not new. Countries have built economic partnerships to facilitate these movements over many centuries. But the term gained popularity after the Cold War in the early 1990s, as these cooperative arrangements shaped modern everyday life. The innovations that reduce transaction costs (better transportation and communication technologies) have always had a strong accelerating impact on global integration. A look back, also suggests that, the process of economic integration is a noncontinuous one. It is often a process of waves that occur when new technologies are widely adopted around the world. Similarly, trade and investment liberalization is being negotiated and implemented in rounds. The two developments, technological change and liberalization, can be mutually reinforcing and create particularly noticeable waves of globalization. The current wave of globalization is driven by major technological breakthroughs in transportation and communication technologies (notably the Internet, mobile telephone technology, and just-in-time systems) in tandem with various efforts to liberalize international trade and investment flows (FAO, 2004).

Food systems are social-ecological systems, formed of biophysical and social factors linked through feedback mechanisms (Berkes et al., 2003; Ericksen, 2008b). They comprise, at a minimum, the activities involved in agriculture and food production, processing and packaging, distribution and retail, and consumption (Ericksen, 2008a). These activities encompass social, cultural, economic, political, institutional, and environmental processes (Tendall et al., 2015).

A food system can be more broadly conceived as including the determinants and outcomes of its activities. The determinants describe the biogeophysical as well as the social, economic, and political environments that determine how food system

Sustainability of the Food System. https://doi.org/10.1016/B978-0-12-818293-2.00010-0

activities are performed (food system drivers). Food system activities and outcomes eventually result in processes that feed back to environmental and socioeconomic drivers (Ericksen, 2008a; FAO, 2008a). These complex interactions and their implications need to be accounted in the design and implementation of effective policy and management interventions. Such interventions thus cannot be treated as isolated changes in one part of the food system (Pinstrup-Andersen and Watson, 2011).

The wide-ranging effects of globalization on food systems are complex and politically charged. As with major technological advances, globalization benefits society as a whole, while harming certain groups. Globalization of technologies faces many interlinked challenges, which jeopardize the sustainability of the food system, such as changing climate conditions, deteriorating natural resources, increasing power imbalances in the agrifood value chain, and changing demographics and dietary habits. The aim of this chapter is to understand the relative costs and benefits of technologies globalization in the food system, to pave the way for alleviating problems while sustaining the wider payoffs and developing strategies toward the sustainability.

2. Technological development and sustainability as components of well-being indicators

Activities of any population group are aimed at ensuring their survival and happiness. It is clear that human well-being is affected positively by technological innovations and negatively by inequality and environmental damages.

The gross domestic product (GDP) is still used as the primary measure of societal progress by every nation in the world. It is an economic indicator that reflects the monetary value of all final goods and services produced by a country or region in a given period of time, usually 1 year. It is used to measure the wealth of a country.

The technological developments of the past decades have meant an increase in satisfaction with a reduction in the price needed to obtain it. But many of the improvements in the quality of life that entails the increase in productivity are not necessarily reflected in an indicator such as GDP. In the same way, this indicator does not reflect the environmental damages derived from a massive use of resources.

A measure of economic well-being should start from income and consumption, taking into account wealth, the perspective of households, income distribution, the extent of inequalities, and the importance of nonmarket activities (Stiglitz et al., 2009). We need better measures of societal well-being that consider the improvements of global technological innovations and, at the same time, the local and sustainable actions aimed at improving the welfare and promotion of life in any population group regardless of size or level of development.

Table 10.1 shows some of the many indicators that have been proposed. Genuine progress indicator (GPI) starts with personal consumption expenditures, a large component of GDP, and adjusts it using 24 different environmental, social, and

Table 10.1 Economic indicators alternatives to the gross domestic product.

Indicator (Reference)	Components Considered	Objective or Advantage
Genuine progress indicator (GPI) Kubiszewski (2019)	Built capital–related components: Personal consumption expenditures, income distribution, personal consumption adjusted for income inequality, services of household capital, services highways and street, cost of consumer durables. Human capital–related components: Value of household labor, value of volunteer work, loss of leisure time, cost of commuting, cost of automobile accidents. Social capital–related components: Cost of crime, cost of family breakdown, cost of underemployment. Natural capital: Cost of household pollution abatement, cost of water pollution, cost of air pollution, cost of noise pollution, loss of wetlands, loss of farmland, depletion of nonrenewable resources, long-term environmental damage, cost of ozone depletion, loss of forest cover.	Allow to estimate net "economic welfare" rather than merely income.
Gross City Development Index (GCD-Index) Ruiz Estrada and Donghyun (2019)	Economic and finance components: Production and consumption of goods and services, income distribution, savings ratio, public and private investment, inflation, and banking. Social components: social protection coverage. Politics and law indicators. Technological components. Environment factors.	The objective of the GCD-Index is to offer policymakers a new analytical tool to assess urban development from a multidimensional perspective. The GCD-Index is a flexible and straightforward indicator that can be applied to analyze the development of any city.

Continued

Table 10.1 Economic indicators alternatives to the gross domestic product.—*cont'd*

Indicator (Reference)	Components Considered	Objective or Advantage
	Population components: Labor, education and training, immigration and migration, and unemployment. Infrastructure and housing factors: Real estate prices and transactions. Income and poverty indicators in formal and informal sectors. Public sector indicators: public transportation, security, health, welfare programs, and taxation. Others factors: historical, customs, habits, religion, values, and anthropological.	
Index of Sustainable Economic Welfare (ISEW) Daly and Cobb (1990)	Include components related to free trade, population, land use, agriculture, industry, labor, taxes, and national security.	Aims to be a proxy of the level of sustainable welfare enjoyed by a population. Directed to an optimal size of community.
Sustainable Welfare Index (SWI) Armiento (2018)	Private consumption expenditure, welfare losses due to income inequality, services provided by nonpaid domestic work, services provided by durable goods, expenditure for durable goods, public expenditure in health and education, R&D expenditure, cost of vehicle accidents, cost of commuting, cost of noise pollution, cost of water pollution, cost of air pollution, cost of urbanization, depletion of nonrenewable resources, social cost of annual carbon dioxide emissions, net fixed capital formation, variation of the net international investment position.	SWI is a flow-oriented indicator that avoids the cumulative accounting of environmental externalities often associated with the use of stock values (e.g., farmlands, wetlands, etc.) and includes certain investments contributing to enhancing future welfare.
National/Regional Welfare Index (NWI/RWI) Held et al. (2018)	Index of Income Inequality (Gini-Index), private consumption weighted by Index of Income Inequality, value of household work, value	The National/Regional Welfare Index shares the ISEW/GPI approach

Table 10.1 Economic indicators alternatives to the gross domestic product.—*cont'd*

Indicator (Reference)	Components Considered	Objective or Advantage
	of voluntary work, nondefensive public expenditure on healthcare and education, net value of the costs and benefits of consumer durables, costs of commuting, costs of traffic accidents, costs of crime, costs of alcohol, tobacco and drug abuse, social expenditure to compensate for negative environmental impacts (costs of avoidance and repair), damage costs of water pollution, damage costs of soil degradation, damage costs of air pollution, damage costs of noise, value of increase/loss of ecosystems, value of increase/loss of agricultural areas, replacement costs due to the consumption of nonrenewable energy resources, damage costs of yearly greenhouse gas emissions, costs of the use of nuclear energy.	

economic components to estimate net "economic welfare" rather than merely income (Kubiszewski, 2019). Alternative measures of economic welfare, such as the Index of Sustainable Economic Welfare (ISEW) and the National Welfare Index (NWI), can be useful tools to help chart the course and assess progress along it if appropriately understood and used (Bleysa and Whitby, 2015).

Although the application of the new indicators has many barriers and disadvantages, these are not insurmountable.

3. The management of food system, technification, and biodiversity

Biodiversity for food and agriculture is the subset of biodiversity that contributes in one way or another to agriculture and food production. It includes the domesticated plants and animals raised in crop, livestock, forest and aquaculture systems, harvested forest and aquatic species, the wild relatives of domesticated species, other wild species harvested for food and other products, and what is known as "associated biodiversity," the vast range of organisms that live in and around food and agricultural production systems, sustaining them and contributing to their output. It is indispensable to food security, sustainable development, and the supply of many vital ecosystem services. It is a key resource in efforts to increase food production while limiting negative impacts on the environment (FAO, 2019). The sustainable use of the components of biodiversity is defined as the use of components of biological diversity in a way and at a rate that does not lead to the long-term decline of biological diversity, thereby maintaining its potential to meet the needs and aspirations of present and future generations (FAO, 2019).

A high mechanization and intensive production systems have soared under the pressure to increase yield. The foundation of our food systems is being determined, often, at least in part, because of the impact of management practices and land-use changes associated with food and agriculture (FAO, 2019). This has been further facilitated by the continuous concentration of the food industry into bigger entities and the economies of scale to reduce production costs.

This has brought negative effects on biodiversity and, consequently, on the sustainability of the food system. Beyond the biodiversity and environmental effects of the use of a limited number of high-yield species or varieties in intensive farming, these crops themselves are also more susceptible to disease and pests. Fertilizers must be supplied regularly to ensure the necessary high crop yield and the supply, and access to these is also finite and complex. Take phosphate (P) as an example; the amount of accessible high-quality phosphate rock is low and diminishing fast; risks to animal and human health are associated with the use of low-quality P fertilizers containing worrying levels of heavy metals such as arsenic, cadmium, and lead (Jiao et al., 2012). Similarly, the increasing use of antimicrobials in animal farming to prevent the spread of diseases has led to the development of resistant animal and zoonotic pathogens. This resistance may even compromise the effectiveness of antibiotic treatments for different human diseases (ECDC/EFSA/EMA, 2015).

An inefficient management of food system has social and economic negative impacts on sustainability, and this effect is enhanced by the globalization of technologies.

Although existing practices and new technologies are already being combined to promote sustainable intensification in developed countries, regional food systems based on a circular economy and self-sufficiency appear more vulnerable in this regard. This niche reality may indeed grow further, triggered by factors such as rising food prices or trends for regional or organic products (Mylonaa et al., 2018).

At the level of the manufacturing food industries, the increase in production through the sustainable use of technology implies to analyze factors influencing sustainable manufacturing. It is important to have a meaningful sustainability index through which manufacturing industries can compare among themselves and measure internal improvement. The focus would be put on nonhazardous waste, and the indicators of the index would be selected with respect to energy efficiency, workers' health and safety, and waste reduction potential (Hasan et al., 2017). Many efforts have been made in this regard. Many authors have tried to obtain an index that considers economic, environmental, and social aspects in a measure of the sustainability of a production plant. However, although there are individual indicators for each of the aspects, the integration of all factors into a single measure is extremely complicated, especially due to the difficulty of assigning a specific weight to each of the parts (Lucato et al., 2018).

4. The nutrition in a sustainable food system

The sustainable use of technology both at the production level and in processing companies involves obtaining nutrient-rich foods and sustainable diets. Additionally, these foods should be available to the entire population and especially to the most vulnerable groups.

Obtaining nutrient-rich foods and sustainable diets will improve malnutrition problems caused both by hunger or lack of food and by overfeeding with low-quality food. In 2017, nearly 151 million children under 5 years of age suffered from stunted growth, whereas 50 million suffered from wasting (a low weight-for-height ratio). Over 38 million children under five were estimated to be overweight and more than 672 million adults to be suffering from obesity (FAO, 2019). A globalized economy focused on maximizing benefits has empowered the big companies, contributing to these data.

The concept of sustainable diets combines the challenges of creating a food system that supplies healthy diets for a growing population while reducing its environmental impacts and staying within planetary boundaries. Sustainable diets are context specific. This highlights the need for context-specific strategies that balance environmental impacts between global and regional scales.

Energy-balanced, predominantly plant-based dietary patterns that are in line with the current evidence on healthy eating can lead to reductions in environmental impacts in high-income and middle-income countries, while improving nutrient levels and reducing diet-related premature mortality in all regions (FAO, 2019). The high consumption of meat has been identified in the most developed countries as one of the main factors responsible for the nonsustainability of diets. Meat production can be an important source of nitrogen, phosphorus, and other pollutants and affects biodiversity—in particular, through land conversion to pasture and arable feed crops. Additionally, high intakes of processed meat are associated with health problems such as cardiovascular disease or colon cancer (Godfrayl et al., 2018).

In low-income countries, adoption of healthier diets can increase demand for environmental resources, in part due to inefficient production systems and baseline diets high in staple crops. Sustainability requires the production increase by means of more efficient production systems: changes in management, technological improvements, and the introduction of new varieties and alternative foods. Some successfully alternatives in specific regions are presented in the following.

Wild foods are food products obtained from nondomesticated species. A wide variety of wild foods, including fruits, leafy vegetables, woody foliage, bulbs and tubers, cereals and grains, nuts and kernels, saps and gums (eaten or used to make drinks), mushrooms, terrestrial invertebrates (insects, snails, etc.), honey, birds eggs, fish, shellfish, and meat from small and large vertebrates, contribute to the diets of large numbers of people, particularly in developing countries (Bharucha and Pretty, 2010).

In the south western part of Ethiopia, a project introduced the use of improved fruit and vegetable varieties, along with practices such as the use of organic manure and integrated pest management, and resulted in a 60% increase in crop yields and a 70% improvement in nutrition in the areas targeted.

The technological development reached at agricultural and industrial level is more than enough to get nutrient-rich foods. But finding effective combinations of policies and approaches that consider local characteristics will be essential for successfully upscaling initiatives and achieving reductions in the health and environmental burden at the population level and globally (Springmann et al., 2018).

The world population is predicted to increase to almost 9.8 billion by 2050 (United Nations, 2017). It means that food supplies and their nutritional quality will need to increase substantially over the coming years and decades (Foley et al., 2011). It has been estimated that global food production needs to be increased by 50% by 2050 (United Nations, 2017). The measures need to be sustainable.

5. The information and communication technologies for a sustainable food system development

As defined by the Agriculture and Rural Development division of the World Bank, an information and communication technology (ICT) is any device, tool, or application that permits the exchange or collection of data through interaction or transmission. The application of digital technologies requires investment and facilities that favor larger companies, farmers with larger fields, and powerful distribution companies and are unavailable to the majority of actors within the food system. However, the ICTs have gained traction even in impoverished regions. The increases in their affordability, accessibility, and adaptability have resulted in their use even within rural homesteads relying on agriculture. New, small devices, infrastructure, and especially applications have proliferated. In some cases, governmental initiatives together with private companies have successfully promoted agricultural and rural development and brought about social and economic changes in developing areas (Swanson and Rajalahti, 2010).

In food system, ICTs have played a role, mostly in supply chain management, but are increasingly being included in farm and land management and even into food manufacturing itself.

In agricultural systems, ICTs can be used for crop management. Together with sensors and measurement tools, ICTs can help to determine the best conditions to add fertilizers or irrigate to the right extent where necessary or to measure the accuracy of a pesticide treatment or even to determine the optimum conditions to cultivate food products (in some cases automatically) (Mangado et al., 2013; Asimopoulus et al., 2013). Moreover, ICTs can be used to prevent or repair environmental damage in natural or cultivated environments (Galiones et al., 2013) and can help to save endangered species (Hunter et al., 2013). On the other hand, people that are engaged in crop and livestock sectors always sought information from each other, and ICTs can facilitate this flow of information and communication between them, and even it is possible to expand the communication to other partners such as experts or local governments to establish and make more feasible rural development plans (Oladele, 2015; Barakade et al., 2011).

The application of ICTs in food industry is not new; their applications have contributed clearly to the safety and nutritional characteristics of the food products by an improvement in the processes control and their optimization (e.g., traceability). The use of ICTs can also help to increase the efficiency of the processes by improving the use of raw materials and decreasing the losses and waste produced (e.g., lean manufacturing).

Related to distribution, marketing, and consumers, ICTs have a crucial role in increasing the companies' transparence, strengthening their relation with consumer, and raising consumer awareness on production, nutritional level, and sustainability of the food products. The latest report on climate change and land of the IPCC, the international panel of experts that advises the United Nations Organization (UN), has declared the need to change the food model to fight the climate crisis. On this occasion, the alert is addressed not only to the industry but also to consumers: "The change in the diet can have large-scale environmental benefits that are not only attainable by the producers" (IPCC, 2019). In this case, ICTs have the possibility to join holistic strategies targeted at educating or informing individual choices and behavior such as ecolabels on food products or awareness campaigns about sustainable resource use (Davis, 2014).

Digital technologies may hold the key to the successful coordination of a more sustainable food system. The world now faces a choice about how to apply technological innovation to the food supply system: to create a global food system in which technologies are applied to increase yields or create new forms of bio-technical food systems, or to create a series of federated local/regional food systems in which smaller-scale local food systems are created and supported through technology. Each of these approaches will need to apply digital technologies to balance the needs of growing populations and scarce natural resources with food security (Berti and Mulligan, 2015).

6. Genetic and biotechnologies

A wide variety of genetic technologies can potentially provide partial or full solutions to the issue of developing sustainable food production systems. To be implemented fully, most of these technologies must be underpinned by information technology. These range from simple issues of food with less land, water, and other inputs, managing wild and farmed populations properly. Each environment and growing system needs a different collection of genes. Increasing genetic knowledge will allow to match species with the most appropriated environment or growing conditions (high- and low-input systems, optimal or marginal environments, etc.).

It is important to ensure technologies are usable and appropriate to smallholders. Partnership and cooperation are necessary between actors (civil society organizations, academia, private sectors, governments, farmers, etc.) for breed development, testing, and wider dissemination of plants and animals with greater resource use efficiency (Bartley and Belanger, 2016). As each technology, if genetic technologies are not properly used, the consequences will be negative. For example, the massive use of one variety promoted by economic interests can suppose an important loose of biodiversity. Biotechnology is not a magic bullet but must be used in conjunction with other aspects of food production. Moreover, efforts are needed to convince consumers and influence policies in favor of their development and application. Communications must be well managed in presenting the facts, especially around food safety.

7. Conclusions

It is evident that human well-being is affected positively by technological innovations and negatively by inequality and environmental damages. Technological development has reached each and every link in the food system. It is clear that technological advances have allowed obtaining more food, safer and healthier than ever.

However, the inappropriate use of technology in an increasingly globalized world in which economic issues prevail over social and environmental ones has brought negative consequences; the depletion of natural resources, the mass production of gases, pollutants and waste, and the increase of social inequalities with problems of hunger and distribution of food of low nutritional value have become a worldwide real concern. The food production system that currently dominates global production has serious negative environmental impacts and is increasingly regarded as unsustainable in a number of respects.

Although it is possible to configure measures and strategies to reverse this trend, the wide-ranging effects of globalization on food systems are complex and politically charged. It is necessary to consider environmental and social aspects linked to technological development as a priority and establish sustainable measures tailored to specific contexts.

Acknowledgements

Authors acknowledge the financial support from Spanish Ministry of Science, Innovation and Universities with the postdoctoral grant Juan de la Cierva Incorporación (IJCI-2016-29679) and with the project AICO/2017/′049 from Generalitat Valenciana.

References

Armiento, M., 2018. The sustainable welfare index: towards a threshold effect for Italy. Ecological Economics 152, 296–309. ISSN 0921-8009.

Asimopoulos, N., Parisses, C., Smyrnaios, A., Germanidis, N., 2013. Autonomous vehicle for saffron harvesting. Procedia Technology 8, 175–182.

Barakade, A.J., Lokhande, T.N., Ubale, P.P., 2011. Impact of globalization information technology in agriculture. International Referred Research Journal 2 (25), 54–57.

Bartley, D., Belanger, J., 2016. Improving productivity through enhanced resource use efficiency. In: Ruane, J., Dargie, J.D., Daly, C. (Eds.), Proceedings of the FAO International Symposium on the Role of Agricultural Biotechnologies in Sustainable Food Systems and Nutrition, pp. 127–129 (Rome, Italy).

Berkes, F., Colding, J., Folke, C., 2003. Navigating Social – Ecological Systems. Building Resilience for Complexity and Change. Cambridge University Press, New York, NY.

Berti, G., Mulligan, C., 2015. Industry Transformation – Horizon Scan: ICT & the Future of Food. Research Report. Ericsson & Imperial College of London, London.

Bharucha, Z., Pretty, J., 2010. The roles and values of wild foods in agricultural systems. Philosophical Transactions of the Royal Society B 365, 2913–2926.

Bleysa, B., Alistair Whitby, A., 2015. Barriers and opportunities for alternative measures of economic welfare. Ecological Economics 117, 162–172.

Daly, H.E., Cobb Jr., J.B., 1990. For the Common Good: Redirecting the Economy toward Community, the Environment, and a Sustainable Future. Beacon Press, Boston, MA, p. 482. ISBN: 0-8070-3702-3.

Davies, A.R., 2014. Co-creating sustainable eating futures: technology, ICT and citizen-consumer ambivalence. Futures 62, 181–193.

ECDC; EFSA and EMA, 2015. ECDC/EFSA/EMA first joint report on the integrated analysis of the consumption of antimicrobial agents and occurrence of antimicrobial resistance in bacteria from humans and food-producing animals. EFSA Journal 13 (1), 4006.

Ericksen, P.J., 2008a. Conceptualizing food systems for global environmental change research. Global Environmental Change 18 (1), 234–245.

Ericksen, P.J., 2008b. What is the vulnerability of a food system to global environmental change? Ecology and Sociology 13 (2C7–14).

FAO, 2019. The state of the world's biodiversity for food and agriculture. In: Bélanger, J., Pilling, D. (Eds.), FAO Commission on Genetic Resources for Food and Agriculture Assessments, p. 572. Rome.

Foley, J.A., Ramankutty, N., Brauman, K.A., Cassidy, E.S., Gerber, J.S., Johnston, M., Mueller, N.D., O'Connell, C., Ray, D.K., West, P.C., Balzer, C., Bennett, E.M., Carpenter, S.R., Hill, J., Monfreda, C., Polasky, S., Rockström, J., Sheehan, J., Siebert, S., Tilman, D., Zaks, D.P.M., 2011. Solutions for a cultivated planet. Nature 478, 337–342.

Food and Agriculture Organization of the United Nations (FAO), 2004. Globalization of Food Systems in Developing Countries: Impact on Food Security and Nutrition. Italy, Rome.

Food and Agriculture Organization of the United Nations (FAO), 2008a. Climate Change and Food Security: A Framework Document. Italy, Rome.

Galionis, D., Charou, E., Stefouli, M., Kabassi, K., Martinis, A., 2013. Mapping burnt forest areas and monitoring their regeneration using MERIS images. Procedia Technology 8, 245–254.

Godfrayl, H.C.J., Aveyard, P., Garnett, T., Hall, J.W., Key, T.J., Lorimer, J., Pierrehumbert, R.T., Scarborough, P., Springmann, M., Jebb, S.A., 2018. Meat consumption, health, and the environment. Review. Science 361 (6399).

Hasan, A., Sarwar, J., Anoman, H., Abdelbaqi, S., 2017. Yearly energy performance of a photovoltaic-phase change material (PV-PCM) system in hot climate. Solar Energy 146, 417–429.

Held, B., Rodenhäuser, D., Diefenbacher, H., Zieschank, R., 2018. The national and regional welfare index (NWI/RWI): redefining progress in Germany. Ecological Economics 145, 391–400.

Hunter, J., Gebers, A., Dettrick, T., 2013. Automating online reef reports cards – linking land management practices to water quality and coral reef ecosystem health. Procedia Technology 8, 183–191.

Intergovernmental Panel on Climate Change (IPCC), 2019. Climate Change and Land. An IPCC Special Report on Climate Change, Desertification, Land Degradation, Sustainable Land Management, Food Security and Greenhouse Gas Fluxes in Terrestrial Ecosystems.

Jiao, W., Chen, W., Chang, A.C., Page, A.L., 2012. Environmental risks of trace elements associated with long-term phosphate fertilizers applications: a review. Environmental Pollution 168, 44–53.

Kubiszewski, I., 2019. The genuine progress indicator: a measure of net economic welfare. In: Brian Fath, Encyclopedia of Ecology, second ed. Elsevier, pp. 327–335. ISBN 9780444641304.

Lucato, W.C., Da Silva Santos, J.C., Tadeu Pacchini, A.P., 2018. Measuring the sustainability of a manufacturing process: a conceptual framework. Sustainability 10, 81.

Mangado, J., Arazuri, S., Arnal, P., Jarén, C., López, A., 2013. Measuring the accuracy of a pesticide treatment by an image analyzer. Procedia Technology 8, 498–502.

Mylonaa, K., Maragkoudakisa, P., Mikob, L., Bocka, A.K., Wollgasta, J., Caldeiraa, S., Ulbertha, F., 2018. Viewpoint: future of food safety and nutrition – seeking win-wins, coping with trade-offs. Food Policy 74, 143–146.

Oladele, O.I., 2015. Effect of information communication technology (ICT) on agricultural information access among extension officers in north west province South Africa. South African Journal of Agricultural Extension 43 (2), 30–41.

Peterson Institute for International Economics (2019). https://www.piie.com/microsites/globalization/what-is-globalization.html Accessed September 2019.

Pinstrup-Andersen, P., Watson, D.D., 2011. Food Policy for Developing Countries. Cornell University Press, Ithaca, NY.

Ruiz Estrada, M.A., Donghyun Park, D., 2019. The application of the gross city development index (GCD-Index) in Tokyo, Japan. Economic Analysis and Policy 62, 37–46.

Springmann, M., Wiebe, K., Mason-D'Croz, D., Sulser, T.B., Rayner, M., Scarborough, P., 2018. Health and nutritional aspects of sustainable diet strategies and their association with environmental impacts: a global modelling analysis with country-level detail. Lancet Planet Health 2, e451–e461.

Stiglitz, J., Sen, A., Fitoussi, J.P., 2009. Report by the Commission on the Measurement of Economic Performance and Social Progress. Available at: http://www.stiglitz-senfitoussi.fr/documents/rapport_anglais.pdf.

Swanson, B.E., Rajalahti, R., 2010. Strengthening Agricultural Extension and Advisory Systems: Procedures for Assessing, Transforming and Evaluating Extension Systems. The

International Bank for Reconstruction and Development/The World Bank, Washington, pp. 98–127.

Tendall, D.M., Joerin, J., Kopainsky, B., Edwards, P., Shrek, A., Le, Q.B., Kruetli, P., Grant, M., Six, J., 2015. Food system resilience: defining the concept. Global Food Security 17–23.

United Nations, 2017. The Sustainable Development Goals Report.

Index

'*Note*: Page numbers followed by "f" indicate figures and "t" indicate tables.'

L

M

N

R

S

T

U

FSC
www.fsc.org
MIX
Papier | Fördert
gute Waldnutzung
FSC® C083411

Kolibri 360 GmbH
Ferdinand-Jühlke-Straße 7
99095 Erfurt, Deutschland
produktsicherheit@kolibri360.de